I dedicate this book to
grandparents, relatives, friends, and teachers.

K.R.

The Geography Bee Comprehensive U.S. Reference Guide

ISBN-10: 1548589314

ISBN-13: 978-1548589318

Printed in the United States

Font Set in Maiandra GD, Arabic Typesetting, Candara, Calibri, Ebrima

Summary: This is a book of geographic questions and vital information designed to prepare students in grades 4-8 competing in the School, State, and National Geographic Bees, Junior/Senior NSF Geography Bees, U.S. Geography Olympiad, and International Geography Bee.

Design and Text by Keshav Ramesh

Cover Illustration by Keshav Ramesh

The Geography Bee Comprehensive U.S. Reference Guide

by Keshav Ramesh

The Geography Bee Ultimate Preparation Guide and A Competitor's Compendium to the Geography Bee:

This young genius has come up with a great guide! His questions follow the various topics that come up on the bee and simulate the types of questions that come.

A must-buy for the Geography Bee contestant!

- **Karan Menon,** 2015 NGB Champion, 2015 U.S. Geography Olympiad JV Champion, and 2017 National U.S. International Geography Bee Third Place Winner

The Geography Bee Ultimate Preparation Guide:

These questions really helped me in my geography endeavors. Keshav is a great author who made many great questions. The content is very well organized and has many different tips and questions. The questions are easy to study from and can help you become smarter in no time!

- **Rohit Gunda,** 2016 Connecticut State Geographic Bee Third Place Winner

The Geography Bee Ultimate Preparation Guide and A Competitor's Compendium to the Geography Bee:

Your books are so useful! I got 3rd at the NJ state bee using just them and travel videos and an atlas. I encourage you to check out books written by Keshav Ramesh, which each has specific sections designed for different levels of the bee. I really appreciated Keshav's book, and it taught me a lot of things. It's a very good book.

-**Ken Mitchell,** 2017 New Jersey State Geographic Bee Third Place Winner

Table of Contents

Tips, Tricks, and How to Prepare for the Geography Bee......8

Ultimate Preparation Guide U.S. Questions 27

Competitor's Compendium U.S. Questions..........................43

Quintessential Questionnaire U.S. Questions.......................57

United States State Fact Files 72

ALABAMA..72

ALASKA 74

ARIZONA...76

ARKANSAS 78

CALIFORNIA 80

COLORADO..84

CONNECTICUT...86

DELAWARE..88

FLORIDA 89

GEORGIA 91

HAWAII 93

IDAHO 95

ILLINOIS 97

INDIANA 98

IOWA...99

KANSAS 100

KENTUCKY 101

LOUISIANA 102

MAINE 104

MARYLAND 106

MASSACHUSETTS 108

MICHIGAN 110

MINNESOTA 112

MISSISSIPPI..114

MISSOURI 115

MONTANA..117

NEBRASKA 119

NEVADA..121

NEW HAMPSHIRE..123

NEW JERSEY 125

NEW MEXICO 127

NEW YORK 129

NORTH CAROLINA..131

NORTH DAKOTA...133

OHIO 134

OKLAHOMA...136

OREGON 138

PENNSYLVANIA 140

RHODE ISLAND 142

SOUTH CAROLINA 143

SOUTH DAKOTA..145

TENNESSEE......147
TEXAS......149
UTAH......151
VERMONT......153
VIRGINIA......154
WASHINGTON......156
WEST VIRGINIA......158
WISCONSIN......160
WYOMING......162
United States Territory Fact Files......164
PUERTO RICO......164
U.S. VIRGIN ISLANDS......166
AMERICAN SAMOA......167
GUAM......168
NORTHERN MARIANA ISLANDS......169
WASHINGTON, D.C. (DISTRICT OF COLUMBIA)......170
Geographic Extremes of the United States......172
About the Author......176
Bibliography......177

Tips, Tricks, and How to Prepare for the Geography Bee

*Note that this section will help you know what to study for U.S. Geography, World Geography, and the other areas of geography (physical, cultural, economic, etc.).

What materials should I have when studying geography?

When studying for the geography bee, I would recommend keeping an atlas, detailed political map, and detailed physical map near you. These three are the most vital sources you need to help you achieve success in the National Geographic Bee.

From looking at a detailed political map, you can find cities and countries from around the world. Territories, islands, and dependencies will be included. You can find oceans, seas, lakes, straits, gulfs, bays, and even rivers. These will be emphasized in a physical map.

There are also smaller versions of political maps – for example, ones showing just South Asia or Central America. Archipelagos are scattered across the world, like Indonesia and Japan. A political map will show you the countries in different colors, and they will be labeled.

Political maps will also show you the continents, and the names of countries will be printed much bigger than the cities.

A physical map will also help as well, to identify the landforms (which will be labeled) as well as biomes. Ocean ridges and seamounts can be displayed on physical maps but not necessarily on all.

Different colors will be expressed on the map, as well as ridges to show mountains, or a plain, yellowish color spreading across a certain area to signify a desert.

Also shown on physical maps are bodies of water, plateaus, geographical regions, basins, and other major parts of the Earth's topography.

In detailed physical maps, or ones that show a close-up of a region or country in the world, also have reservoirs, smaller bays/gulfs, highlands, hills, passes, and much more.

Atlases are great for researching thousands of facts just by looking at maps and diagrams, and reading about them. Countries are always featured in atlases, and you'll always find a few chapters about the physical geography of the world.

You will find many thematic maps in atlases, such as those depicting religions, language families, economy, water supply, major agricultural commodities, and more.

I would recommend buying atlases sold by National Geographic, as they, in my opinion, provide the best information.

The National Geographic Atlas of the World, Tenth Edition is good for participants serious about winning the state bees and doing well in the national bees. There is no atlas like this in the world that will have a lot of very detailed information. Although it is expensive, it is worth the price. State winners have been awarded this book in the past.

You should get an almanac, like the *National Geographic Kids Almanac 2018* or another good one recently published from National Geographic with interesting (and sometimes weird!) geography facts.

However, this book gives you a broader outlook on geography, so it is a good reference for your school and sometimes even your state bees.

The *National Geographic Magazine* is especially important, as questions regarding facts in the magazine have been asked.

If you are participating in the 2018 National Geographic Bee, learn the geographical facts from the magazine from its six issues before May (November 2017-April 2018). This applies to 2019 and beyond as well.

You should also get a book about the physical geography of the world, as this is vital to the physical geography section of the National Geographic Bee. The *National Geographic Desk Reference* is probably the best option for you when looking at physical geography.

In general, what is in this book?

The Geography Bee Comprehensive U.S. Reference Guide has **nearly 10,000 facts, locations, landforms, sites, and landmarks to help you in all United States rounds of the National Geographic Bee!** The information is separated alphabetically by state, in a structured format:

- Basic Information
- Physical Geography and Landforms
- Protected Areas
- Other Landmarks (manmade or protected)
- Miscellaneous Information (dams and international airports)

There are also subtopics in some of the geographical topic chapters (Physical Geography, Cultural Geography, Economic Geography, etc.).

What is the daily amount of time I should study?

If you want to win the school bee, studying for 30 minutes to 1 hour a day is enough (unless you are in a competitive school for the geography bee for those of you in New Jersey, Florida, California, Washington, Michigan, Virginia,

Maryland, or Texas). For the state level bee, I would recommend studying 2 to 3 hours a day.

If you are at the level of the National Bee Preliminaries, 3 to 4 hours is what I would recommend. Those aspiring to achieve a place in the top ten should definitely amount to more than 4 hours, if not 5-7.

This competition, like others, is challenging. Hours of dedication to geography is vital to your success in the state and national bees.

Is this book only for competitors in the National Geographic Bee?

This book is not only for people competing in the National Geographic Bee, but for other competitions as well.

If you are participating in the finals of the **North South Foundation's Junior/Senior Geography Bees**, this is a good book for you. Using this structured information to your advantage will help you get far ahead in the competition.

This book contains questions not only written for the National Geographic Bee, but also for the **United States**

Geography Olympiad (USGO). The USGO starts regionally as a National Qualifying Exam and any competitor must score above the national median score and/or in the top half at the regional level.

The **National USGO Championships** are held along with the **National History Bee and Bowl,** and the U.S. International Geography Bee in Arlington, Virginia annually in April. This Olympiad selects the top four individuals in the Varsity Division to represent the United States at the **International Geography Olympiad (iGeo)**, held in a different city around the world annually.

This book has also been written for students preparing for the **International Geography Bee (IGB)**, an intense geography competition created in 2017 with a regional National Qualifying Exam and fast quiz bowl rounds at the national level.

The U.S. national championships of the International Geography Bee is held in conjunction with the USGO and the **National History Bee and Bowl** annually in Arlington, Virginia in April. The top four individuals in the JV Division represent the United States region at the **Junior Varsity IGB World Championships.** The top four individuals in the

Varsity division represent the United States region at the **Varsity IGB World Championships.**

As a list, this book is a great resource for these competitions:

- National Geographic Bee (NGB)
- North South Foundation Senior Geography Bee (SGB)
- North South Foundation Junior Geography Bee (JGB)
- United States Geography Olympiad (USGO)
- International Geography Bee (IGB)

What should I study for the state and national competitions of the National Geographic Bee?

For the state bee, knowledge of borders, locations, and the United States is important. Be sure to know your national parks, national forests, and national monuments. You should be well versed in the major cities of each state and country.

Be sure to follow National Geographic's Instagram and Twitter for more geographic information. **In the 2016 State Geographic Bees, questions regarding photos from**

National Geographic's Instagram account were posed to the competitors in the final rounds.

If you reach the finals of the state bee, your atlas should be your greatest ally. You should have a complete mastery or near mastery of thousands and thousands of locations – however, this should not be achieved through memorization, but through constant review. It may sound difficult, but it gets much easier after you practice and practice, at least three hours a day.

Although knowing thousands of locations sounds like a lot, you probably know the name and location of every single country, some major cities, major rivers, major lakes, major mountain ranges, notable islands, deserts, oceans, and seas if you are an experienced geography bee participant. This by itself is a lot of information.

Be thorough with your atlases, and use the National Geographic Atlas of the World, Tenth Edition as much as possible.

Create questions, research facts, connect locations, and take notes. Keep rereading what you have read and written down/typed and try to produce questions from

the facts. Integrate the facts and questions as much as possible so that you can remember them.

You should be thorough with tourist attractions and landmarks, as well as river confluences, port cities, UNESCO World Heritage Sites, major exports, current events, physical geography terms, and geographic extremes (such as Kanyakumari, Cape Byron, and Ushuaia).

What do I need to know about each state?

Here's a list of what you need to know about each state to help you prepare for the National Geographic Bee. Remember, some of the things on this list may not apply to certain state:

Basics:

- Location (Region, e.g. New England)
- Capital
- Major Cities (10+ for states with populations over 3,000,000)
- Population (Approximate)

Physical:

- Highest and Lowest Points
- Mountain Ranges, Peaks, and Volcanoes
- Rivers, Deltas, River Mouths/Sources
- Lakes, Reservoirs, and Dams
- Bordering Seas
- Bays and Gulfs
- Straits, Sounds and Inlets
- Plateaus
- Peninsulas, Capes, and Points
- Plains and Basins
- Isthmuses and Spits
- Wetlands, Swamps, and Marshes
- Deserts and Dunes
- Valleys, Grasslands and Prairies
- Canyons and Gorges
- Islands and Archipelagoes
- Major Glaciers
- Other Bodies of Water
- Other Landforms

Political

- Bordering States
- Territories, Dependencies, and Occupied Atolls
- Government Structure/Important and Influential Laws

Cultural

- Languages
- Foods, Art, Music, and Architecture
- Cultural Items/Objects and Symbols

Environmental

- Conservation and Biodiversity
- Biomes and Habitats
- Plants and Animals
- Global Warming/Climate Change
- Environmental Hot Spots
- Natural Disasters

Economic

- Production
- Agricultural Products and Natural Resources
- Port Cities, Seaports, and International Airports

Historical:

- U.S. History
- Colonies
- Treaties and Pacts
- Wars

Landmarks

- National Parks/Preserves, Forests, and Monuments
- National Historic Sites and Historical Parks

- Museums and Zoos
- Buildings and Structures in Different Cities
- Skyscrapers and Towers
- Business and Financial Districts
- Space Centers, Observatories, and National Laboratories

Current:

- United Nations
- Major Nationwide Non-Sporting Events
- Major Nationwide Sporting Events (Super Bowl, U.S. Open, Indianapolis 500)

What are some good websites I can use to prepare?

The website I'd recommend to help you prepare for the competition is www.nationalgeographic.com/geobee. It has great tools to help you study, and also a page where you can play a game called the GeoBee Challenge, where National Geographic gives you 10 new questions every day to prepare with.

National Geographic has also partnered with Kahoot to create geography bee prep games. Below is a list of all websites you should probably use to enhance your knowledge.

Sites Dedicated Exclusively to Geography and the Geography Bee:

www.nationalgeographic.org/geobee

www.geobeeworld.blogspot.com

www.geography.about.com

http://lizardpoint.com/geography/

http://www.sporcle.com/games/category/geography/all

http://nationalgeographic.org/bee/study/play-kahoot/

What are some other online resources I can use?

Quizlet is an online quizzing program that you can use to prepare for the geography bee.

I would also use Socrative, where you can make multiple choice questions with multiple answers (if you choose). There are hundreds, maybe thousands of geography quizzes on Sporcle.

The world's largest geography bee community is on Google+. Look up **GeoBee City** on Google+, and you will find it. GBC is a geography bee community where students from all over the country work on geography-related projects, mock bees, quizzes, maps, and use resources to study. GBC was founded by Karan Menon, the 2015 National Geographic Bee Champion, in 2014 and the community has over 200 members.

In addition to studying for the National Geographic Bee with fellow students, we prepare for the **United States Geography Olympiad** and the **International Geography Bee**, and some of us who are eligible prepare for the **North South Foundation (NSF) Junior and Senior Geography Bees.**

Any questions or concerns?

Any questions you'd like me to add? Still need help with geography? Any errors that need to be fixed? Contact me at keshav.ramesh@gmail.com!

Is there anything else I should know when preparing for the geography bee?

There is!

Let us set the scenario that you were participating in the 2012 State Geographic Bee in fourth grade, up to the 2016 State Geographic Bee in eighth grade. This way, National Geographic has a near-perfect way of guaranteeing that you wouldn't have known some of the questions asked in the 2011 State Geographic Bee finals as you would've been in third grade and most likely not watched the final rounds.

Let us look at a different scenario. Let's say you were participating in the 2013 State Geographic Bee in fourth grade, up to the 2017 State Geographic Bee in eighth grade. Again, wouldn't National Geographic have a near-perfect way of guaranteeing that you wouldn't have known some of the questions asked in the 2012 State Geographic Bee? So instead of creating a set of new final round questions, they could just reuse them.

Certainly, the National Geographic Society would have to create new questions for the state finals – but not all of the questions would necessarily have to be created as of that year.

So for those of you participating in the 2018, 2019, or 2020 State Geographic Bee, I would suggest looking at these videos below now or sometime later, and make sure that you record the questions, test yourself on them, and maybe find some other patterns between the questions of different years:

2013 for 2018:

http://ct-n.com/ctnplayer.asp?odID=8880

2014 for 2019:

https://www.youtube.com/watch?v=rJX_YSSOVU8

2015 for 2020:

https://www.youtube.com/watch?v=Tu3xJLqJIFA

2016 for 2021:

https://www.youtube.com/watch?v=6FuT9nAI9dY

2017 for 2022:

http://ct-n.com/ctnplayer.asp?odID=13916

In the preliminaries round of the 2017 National Geographic Bee, competitors went through what was called the **Explorers Round.**

Each competitor would be asked a question and their answer would have to be a National Geographic explorer from one of two choices given.

This round took away points from many participants, as most had not studied individual National Geographic explorers.

Here are two links you can use to prepare for this round should it ever come up in a future National Geographic Bee:

http://www.nationalgeographic.com/explorers/explorers-a-z/

http://www.nationalgeographic.com/explorers/explorers-category-search/

In the preliminaries round of the 2017 National Geographic Bee, competitors went through what was called the **Ocean in Motion** round.

Recent National Geographic Bees have also increased the number of questions and rounds regarding Oceans, Conservation, Climate Change, and Environmental Science.

Here are some links you can use to prepare for this round should it ever come up in a future National Geographic Bee (and it probably will!). I hope it works!:

https://en.wikipedia.org/wiki/Glossary_of_ecology
https://en.wikipedia.org/wiki/Glossary_of_environmental_s

cience

https://en.wikipedia.org/wiki/Glossary_of_climate_change

https://en.wikipedia.org/wiki/List_of_severe_weather_phe
nomena

Should I use National Geographic's Instagram account to help me study?

Using National Geographic's Instagram account to learn geography is a great idea. (you don't need an Instagram account to do this!).

Many questions are actually based off of pictures National Geographic posts of oceans, animal and plant species, conservation projects, geography, science, and nature. Check out these links!:

https://www.instagram.com/natgeo/

https://www.instagram.com/natgeotravel/

https://www.instagram.com/natgeoyourshot/

https://www.instagram.com/natgeowild/

Good luck and happy studying for the National Geographic Bee!

Ultimate Preparation Guide U.S. Questions

1. The Lewis and Clark Expedition explored which region from 1804 to 1806 – Louisiana Purchase or Southwest?
   Louisiana Purchase

2. Lake Sakakawea is a 200-mile-long (320-kilometer-long) reservoir along which river – Missouri River or Colorado River?
   Missouri River

3. What city, founded by William Penn, was the site of the First and Second Continental Congresses – Baltimore or Philadelphia?
   Philadelphia

4. Which one of the contiguous United States has the northernmost point – Minnesota or Wisconsin?
   Minnesota

5. Which city is more likely to have desert vegetation – Phoenix or Denver?
   Phoenix

6. Which city is located on the western tip of Lake Superior and is an important port for shipping grain and iron ore – Duluth or Chicago?
   Duluth

7. What city in Alabama was the subject of a 1955 and 1956 bus boycott organized by Martin Luther King, Jr., to protest bus segregation – Montgomery or Birmingham?
   Montgomery

8. Which state has more coastal marshland – Louisiana or Kentucky?
   Louisiana

9. Which city is in western Texas and lies directly across the border from the Mexican city of Ciudad Juarez – Amarillo or El Paso?
   El Paso

10. Badlands National Park and Mount Rushmore are west of the Missouri River in which state – Minnesota or South Dakota?
    South Dakota

11. Which northeast state borders Canada – New Hampshire or Massachusetts?
    New Hampshire

12. Cape Cod, known for its beautiful shoreline, is located in which state – New Jersey or Massachusetts?

Massachusetts

13. When it's 1 p.m. in Hawaii, what part of the day is it in the state of New York – midnight or evening?
Evening

14. Which state produces more wood pulp – Georgia or Florida?
Georgia

15. Land bordering the mouths of the James, York, and Rappahannock Rivers is part of the Tidewater region of which state – Virginia or Ohio?
Virginia

16. Poi is a food commonly associated with people native to which state – Hawaii or Alaska?
Hawaii

17. The melaleuca tree is believed to be drying up parts of a large swamp, making it uninhabitable to native plants and animals. This process is occurring in what swampy region – Great Basin, Everglades, or Pine Barrens?
Everglades

18. Tornadoes occur most frequently in which region of the United States – Northwest, Southwest, or Midwest?
Midwest

19. Which state has more national parks – California, Delaware, or Iowa?
California

20. The Constitution State is the state nickname for what state – Connecticut or Maryland?
Connecticut

21. The easternmost point in the contiguous United States is located in what state – Massachusetts or Maine?
Maine

22. El Paso is a city that lies close to Mexico in what state famous for its oil production – Texas or New Mexico?
Texas

23. Mount St. Helens is a volcano in what state – Oregon or Washington?
Washington

24. Miami, Orlando, and Tallahassee are all cities in what state home to the Everglades and Walt Disney World – Louisiana or Florida?
Florida

25. Chicago borders what Great Lake – Lake Michigan or Lake Huron?
Lake Michigan

26. Florida, Texas, Louisiana, Alabama, and what other southeastern state borders the Gulf of Mexico – Mississippi or Georgia?
    Mississippi

27. The Great Miami River and Little Miami River are two rivers in what U.S. State in the Eastern Time Zone – Florida or Ohio?
    Ohio

28. The Golden Gate Bridge is located in what state that is home to the cities of Cupertino and Oakland – New Jersey or California?
    California

29. Cape Canaveral is located in what state – Florida or Massachusetts?
    Florida

30. The port of Mobile is located in what U.S. State – Mississippi or Alabama?
    Alabama

31. In which state is more land used for agriculture – Connecticut or Illinois?
    Illinois

32. Which state borders California on the north – Oregon or Utah?
    Oregon

33. In July 2002, the U.S. Congress voted to authorize Yucca Mountain as a permanent repository for 77,000 tons of nuclear waste. Yucca Mountain is located in which state – Kansas or Nevada?
    Nevada

34. The deep roots of the tamarisk, or salt cedar tree, can deprive native plants of water. Although concentrated in the arid southwest, this tree has spread as far as what state located north of Wyoming – Idaho or Montana?
    Montana

35. Newport, a prominent seaport in colonial times, is now a popular tourist area located in which state – Rhode Island or Delaware?
    Rhode Island

36. Hydrilla, introduced to Florida as an aquarium plant, now clogs many waterways in the United States. This plant, which is spread easily by boats, has been found near Washington, D.C., in which river – Hudson or Potomac River?
    Potomac River

37. Brownsville, an important border city on the Rio Grande, is located in which state – New Mexico or Texas?
    Texas

38. Which state does not have a coastline – New Mexico or New Hampshire?
    New Mexico

39. Which bay has a larger total area – Prudhoe Bay or Chesapeake Bay?
    Chesapeake Bay

40. The Brooks Range is located in what state – Maine or Alaska?
    Alaska

41. Mount Whitney is the highest point in the contiguous United States, located in what state – California or Nevada?
    California

42. The Colorado Plateau, located in the southwestern United States, is between the Rocky Mountains and what else – the Great Basin or Ouachita Mountains?
    The Great Basin

43. The Chilkoot Pass crosses from Alaska into what country – Canada or Mexico?
    Canada

44. The Everglades, a large swamp, can be found in the southern part of what state where Walt Disney World can be found?

Florida

45. The San Joaquin Valley is located between two mountain ranges. These mountain ranges are the Sierra Nevada and what other range – Coast Range or Cascade Range?
Coast Range

46. The Chihuahuan Desert, found in Texas and Mexico, extends into what southwestern U.S. state – Arizona or New Mexico?
New Mexico

47. The United States borders an amazing three oceans. These oceans are the Atlantic, Pacific, and what other ocean including the Beaufort and Chukchi Seas – Arctic or Southern?
Arctic Ocean

48. Puget Sound feeds into the Pacific Ocean. This sound is located in what state whose capital is Olympia and is home to the Seattle Seahawks football team?
Washington

49. The Arkansas River, a tributary of the Mississippi River, has its source in what major mountain range – Rocky Mountains or Appalachian Mountains?
Rocky Mountains

50. The Great Salt Lake is located in what U.S. state home to a large population of Mormons – Utah or Wyoming?
    Utah

51. Lake Okeechobee, on the edge of the Everglades, is located in what U.S. state with the cities of Orlando and Miami – Georgia or Florida?
    Florida

52. Chesapeake Bay is an estuary of what river – Susquehanna River or Allegheny River?
    Susquehanna River

53. Lake Pontchartrain is located on what plain – Great Plains or Gulf Coastal Plain?
    Gulf Coastal Plain

54. Yosemite Falls, located in the Sierra Nevada, is in what region of the United States – southern or western?
    Western

55. Cape Cod and Cape Canaveral are located on what coast of the United States – West or East?
    East coast

56. Howland Island, Jarvis Island, and Wake Island are all territories of the United States located in what ocean – Atlantic Ocean or Pacific Ocean?
    Pacific

57. People from Mexico and what other country constitute the highest number of U.S. immigrants every year – India or China?
India

58. Permafrost in the United States is found in what U.S. state – Montana or Alaska?
Alaska

59. Molokai is one of the main islands of what U.S. state that was the last to join the U.S. – Alaska or Hawaii?
Hawaii

60. Houston is a major port on what gulf – Gulf of Mexico or Gulf of Alaska?
Gulf of Mexico

61. The District of Columbia (Washington, D.C.) is located between Maryland and what other U.S. state – Delaware or Virginia?
Virginia

62. Cape Hatteras can be found off the coast of what U.S. state home to Duke University – North Carolina or South Carolina?
North Carolina

63. Long Island Sound is north of Long Island and south of what U.S. state home to the historical figures of Oliver

Ellsworth and Noah Webster – Connecticut or Rhode Island?
Connecticut

64. Glen Canyon is a hydroelectric dam on what river – Monongahela River or Colorado River?
Colorado River

65. The Gulf of Maine feeds into what ocean – Pacific or Atlantic?
Atlantic

66. Lake Michigan is the only Great Lake entirely in the United States. This lake also borders what U.S. state where the cities of Lansing and Detroit can be found? – Michigan or Illinois?
Michigan

67. Which Great Lake is the smallest in size – Ontario or Erie?
Erie

68. The Saint Lawrence River forms part of the border between the United States and what other country – Mexico or Canada?
Canada

69. The Seward Peninsula is located in what U.S. state whose major cities include Anchorage and Fairbanks – Alaska or Washington?

Alaska

70. Phoenix, a major U.S. city and the capital of Arizona, is on the edge of what southwestern U.S. desert – Sonoran Desert or Chihuahuan Desert?
Sonoran Desert

71. Chicago is the chief port on what lake – Lake Michigan or Lake Superior?
Lake Michigan

72. Cape Flattery is bordered by the Pacific Ocean and what strait – Strait of Juan de Fuca or Bering Strait?
Strait of Juan de Fuca

73. The Aleutian Islands are a group of islands belonging to Alaska that extend across what meridian – 180 degree or 170 degree?
180 degree

74. Hurricane Katrina is famous for wrecking what major city – New Orleans or Mobile?
New Orleans

75. Amarillo is a city in what U.S. state famous for its oil production – Oklahoma or Texas?
Texas

76. The Straits of Florida connect the Atlantic Ocean to what gulf – Gulf of Alaska or Gulf of Mexico?

Gulf of Mexico

77. The Colorado River, emptying out into the Gulf of California, created what major canyon in Arizona – Grand Canyon or Palo Duro Canyon?
    Grand Canyon

78. Hoover Dam can be found on the border between Arizona and what other U.S. state known as the Silver State – Nevada or New Mexico?
    Nevada

79. The longest river in New England is what river – Connecticut River or Hudson River?
    Connecticut River

80. Lake Itasca can be found in what U.S. state – Mississippi or Minnesota?
    Minnesota

81. Hartford, known as the insurance capital of the United States, is the capital of what U.S. state where the first hamburger was served in U.S. history – Connecticut or Rhode Island?
    Connecticut

82. Wisconsin borders what U.S. state to the east – Michigan or Illinois?
    Michigan

83. Louisville and Knoxville are cities in what U.S. state that straddles the world's longest Cave System, Mammoth Cave National Park – Kansas or Kentucky?
    Kentucky

84. Lake Champlain borders New York and what other U.S. state – Vermont or New Hampshire?
    Vermont

85. Redwood National Park can be found in the northern part of what U.S. state – California or Oregon?
    California

86. Marquette and Flint are cities in what U.S. state bordering four of the Great Lakes – Wisconsin or Michigan?
    Michigan

87. International Falls can be found in what U.S. state with the cities of Duluth and Rochester – Minnesota or New York?
    Minnesota

88. Cape Mendocino and Point Conception can be found in what U.S. state famous for the Sierra Nevada, Mojave Desert, and Mt. Whitney – Nevada or California?
    California

89. Which bay is connected to the Atlantic Ocean – Monterey Bay or Chesapeake Bay?

Chesapeake Bay

90. Which U.S. state has a longer border with Mexico – Texas or California?
Texas

91. Which city has a greater risk of experiencing an earthquake – Los Angeles or New York City?
Los Angeles

92. New York is the only U.S. state that borders which lake – Lake Erie or Lake Ontario?
Lake Ontario

93. Which U.S. state does not experience frequent earthquakes – California or Massachusetts?
Massachusetts

94. Cape Cod, Cape Hatteras, and Cape May are along which coast of the United States – West or East?
East

95. Which one of the 50 states shares the longest border with Mexico – New Mexico or Texas?
Texas

96. Which city in southern Nevada saw its population grow by more than eighty percent in the 1990s – Las Vegas or Reno?
Las Vegas

97. Water from the Snake River makes it possible for which mountainous U.S. state to produce more than one-quarter of the nation's potato crop – Idaho or Wyoming?
Idaho

98. Albuquerque, New Mexico, is to the Rio Grande as Bismarck, North Dakota, is to what – Snake River or Missouri River?
Missouri River

99. Which U.S. state capital is farther north – Augusta, Boston, or Hartford?
Augusta

100. Which city is located at a higher elevation – Denver or New Orleans?
Denver

Competitor's Compendium U.S. Questions

1. Havre and Great Falls are cities in what U.S. State bordering North Dakota and Canada?
   Montana

2. The Near Islands belong to what U.S. State?
   Alaska

3. Cupertino can be found in what U.S. State that is home to the Mojave Desert?
   California

4. The majority of French-speaking people in the United States live in what state bordering Arkansas to the north?
   Louisiana

5. Pocatello is a city near the Snake River in what U.S. State?
   Idaho

6. Sault Ste. Marie is at the tip of what peninsula bordering Lake Superior and Lake Huron?

Upper Peninsula

7. Galveston is a city located miles southeast of what major Texan city?
Houston

8. Lake Okeechobee and West Palm Beach are located in what U.S. State?
Florida

9. Coyotes originated in what geographical region of the United States?
Southwest

10. Hawaii, although part of the United States, is geographically part of what region in Oceania?
Polynesia

11. What U.S. State is known as the lightning capital of the United States?
Florida

12. Chrysler Building is located in Manhattan in what U.S. State?
New York

13. Castle Geyser is a famous geographical attraction part of what national park?
Yellowstone National Park

14. Hyperion, the world's tallest living tree, is taller than the Statue of Liberty. It is located in Redwood National Park in what U.S. State known as "The Golden State"?
    California

15. The Aleutian Mountain Range is located on what peninsula in Alaska?
    Aleutian Peninsula

16. Mount St. Helens is a famous peak in what mountain range extending from northern California to Washington?
    Cascade Mountain Range

17. The Colorado Plateau is located between the Cascade Mountains and what basin?
    The Great Basin

18. The Everglades is a swampy region located on what peninsula in the southeastern United States?
    Florida

19. The Chilkoot Pass crosses from the United States into what country?
    Canada

20. The Strait of Juan de Fuca separates the United States from what country?
    Canada

21. The Gulf of Alaska feeds into what ocean?
    Pacific Ocean

22. Lake Pontchartrain, in the southern United States, is the largest lake in what U.S. State?
Louisiana

23. The Rio Grande, a river forming much of the border between the United States and Mexico, has its source in what major mountain range?
Rocky Mountains

24. The Colorado River, which forms the Grand Canyon, empties out into what gulf with the same name as a U.S. State?
Gulf of California

25. Yosemite Falls is located in what major mountain range in California?
Sierra Nevada

26. Glen Canyon is a hydroelectric dam on the Colorado River creating what major lake?
Lake Powell

27. The Seward Peninsula borders the Bering Strait, Bering Sea, and what other sea?
Chukchi Sea

28. The Mississippi River Delta is located in what U.S. State with the cities of New Orleans and Baton Rouge?
Louisiana

29. The Snake River forms Hells Canyon, the deepest gorge in the United States and is a tributary of what river forming much of Oregon's border with Washington?
Columbia River

30. Cape Hatteras is a chain of barrier islands in the Atlantic Ocean off the coast of what U.S. State?
North Carolina

31. Kingman Reef is a territory of the United States in what ocean?
Pacific Ocean

32. Houston is a major port city on what body of water east of Mexico?
Gulf of Mexico

33. Los Angeles is a city on the edge of the Coast Ranges in what U.S. State?
California

34. Cape Flattery can be found at the northwestern tip of what peninsula in Washington?
Olympic Peninsula

35. Phoenix is a city in Arizona on the edge of what major desert?
Sonoran Desert

36. Katmai National Park is located in what U.S. State where you can view the Aurora Borealis?
Alaska

37. The Susquehanna and Allegheny Rivers are located in what U.S. State where the cities of Allentown and Lancaster can be found?
    Pennsylvania

38. Yale University is an Ivy League in what U.S. State bordering Long Island Sound?
    Connecticut

39. The Catskill and Adirondack Mountains are located in what U.S. State?
    New York

40. Mt. Frissell is the highest point in what U.S. State bordering New York and Massachusetts?
    Connecticut

41. Lake Champlain is located in New York and what U.S. State whose capital is Montpelier?
    Vermont

42. The Finger Lakes and the Mohawk River are in what U.S. State whose capital is Albany?
    New York

43. The easternmost point in the United States is located in what U.S. State bordering Quebec and New Brunswick?
    Maine

44. Lake Chocurua is located in the White Mountains in what U.S. State?

New Hampshire

45. Mystic Seaport is located in what U.S. State home to the Housatonic and Naugatuck Rivers?
Connecticut

46. Prime Hook National Wildlife Refuge is located in what U.S. State bordering Delaware Bay?
Delaware

47. Acadia National Park is located in what U.S. State including Mooselookmeguntic Lake and Sebago Lake?
Maine

48. The Chesapeake Bay Bridge is located in what U.S. State whose capital is Annapolis?
Maryland

49. What language is the second most spoken language in the United States?
Spanish

50. The Salton Sea is north of what valley in southern California, west of the Colorado River?
Imperial Valley

51. Mt. Elbert is located in what major mountain range west of the Park and Front Ranges?
Rocky Mountains

52. The Davis Mountains are north of the Rio Grande in what U.S. State?

Texas

53. The Pearl River is located in what U.S. State bordering Alabama?
Mississippi

54. The Yellowstone River, Milk River, and Flathead Lake are located in what U.S. State?
Montana

55. Lake of the Ozarks is north of the Ozark Plateau in what U.S. State straddling part of the Missouri River?
Missouri

56. Isle Royale is an island belonging to what U.S. State bordering Ontario?
Michigan

57. The Des Moines River flows through what U.S. State that is part of the Central Lowland and borders Minnesota and Nebraska?
Iowa

58. The Allegheny Plateau is west of what mountain range through which the Susquehanna River flows?
Appalachian Mountains

59. Long Island Sound is south of what U.S. State bordering Rhode Island?
Connecticut

60. The Toledo Bend Reservoir is on Texas's border with what U.S. State bordering Atchafalaya Bay?
Louisiana

61. The Cumberland Plateau is east of the Cumberland and Tennessee Rivers. This plateau is also west of what major mountain range?
Appalachian Mountains

62. Point Barrow is a city in what U.S. State where the Tanana and Kuskokwim Rivers can be found?
Alaska

63. Lanai is an island in what U.S. State home to the peak of Mauna Kea?
Hawaii

64. The Colorado Plateau encompasses Colorado, Utah, Arizona, and what U.S. State?
New Mexico

65. The Edwards Plateau is located north of the Rio Bravo del Norte in what U.S. State?
Texas

66. Georgian Bay is an inlet of which of the five Great Lakes?
Lake Huron

67. Apalachee Bay is south of what U.S. State straddling part of the Gulf Coastal Plain?
Florida

68. The St. Johns River is located in what U.S. State where Cape Sable can be found?
    Florida

69. The Sangre de Cristo Mountains are located in Colorado and what U.S. State home to the Elephant Butte Reservoir?
    New Mexico

70. The Sacramento Mountains are located in what U.S. State whose capital is Santa Fe?
    New Mexico

71. Moosehead Lake is located in what U.S. State bordering the Bay of Fundy?
    Maine

72. The Green Mountains are to Vermont as the White Mountains are to what?
    New Hampshire

73. What river, named after a U.S. State in this region, is the longest river in New England?
    Connecticut River

74. The Connecticut River forms the border between what two U.S. States?
    Vermont and New Hampshire

75. The Flint Hills are located in what U.S. State home to the Smoky Hill River?
    Kansas

76. The Yazoo River is located in what U.S. State bordering the Gulf of Mexico and Tennessee?
    Mississippi

77. The Sacramento Valley is located in what U.S. State bordering Monterey Bay?
    California

78. The Channel Islands and the Sierra Nevada are in what state?
    California

79. Harney Peak is in the Black Hills in what state west of Minnesota?
    South Dakota

80. The Ruby Mountains are located in what state home to the Humboldt River?
    Nevada

81. The Olympic Mountains are located in what state?
    Washington

82. The Connecticut River empties out into what sound?
    Long Island Sound

83. The Nushagak Peninsula is located north of Bristol Bay in what state?
    Alaska

84. Craters of the Moon National Monument is located in what state?
Idaho

85. Albemarle Sound and Pamlico Sound are on what coast of the United States?
East Coast

86. The Cortez Mountains and Ruby Mountains can be found in what U.S. State bordering Oregon and Idaho?
Nevada

87. Golden Spike National Historic Site is located northeast of the Great Salt Lake in what state?
Utah

88. Arabic, Chinese, English, French, Russian, and what other language are the official language of the United Nations?
Spanish

89. Ban Ki-moon is the current secretary-general of what major world organization?
United Nations

90. The Sonoran Desert is in what state whose highest natural point is Humphreys Peak?
Arizona

91. Lake St. Clair borders what state?
Michigan

92. The Keweenaw Peninsula belongs to what state bordering Lake Michigan?
    Michigan

93. The most populous metropolitan area in the United States is at the mouth of the Hudson River. Name this city.
    New York City

94. Wilkes-Barre is a city in what state bordering New York?
    Pennsylvania

95. The Boston Mountains are south of what plateau?
    Ozark Plateau

96. Haleakala National Park is located on what Hawaiian Island?
    Maui

97. The Trinity Islands belong to what state bordering Canada to the east?
    Alaska

98. The Niagara River cascades over what waterfall also known as Canadian Falls and part of Niagara Falls?
    Horseshoe Falls

99. The Shumagin Islands belong to what state?
    Alaska

100. Murfreesboro is a major city in what state?
    Tennessee

Quintessential Questionnaire U.S. Questions

1. What is the largest city by population in the panhandle of Connecticut, in the southwestern part of the state?
   Stamford

2. Reno is a major city in what U.S. state?
   Nevada

3. Mystic Seaport is located in the southern part of what New England state?
   Connecticut

4. What state is famous for its Cajun culture and borders the Gulf of Mexico?
   Louisiana

5. Name the only national park that in South Carolina.
   Congaree National Park

6. Tampa and St. Petersburg are major cities in what state?
   Florida

7. Padre Island belongs to what state with a panhandle?
   Texas

8. The Aleutian Islands are an archipelago belonging to what Pacific state?
   Alaska

9. The Kittatinny Mountains form the Wallpack Valley in what Mid-Atlantic state?
   New Jersey

10. You can ride a train at the Jelly Belly warehouse near the city of Kenosha in which state?
    Kenosha

11. The Natural Bridge Caverns, the largest known commercial caverns in Texas, are near what major city that is in the southwestern part of the Texas Triangle?
    San Antonio

12. You can create your own candy bar at Hershey's Chocolate World near Harrisburg in which state?
    Pennsylvania

13. Stonehenge II, a model of the original Stonehenge in the United Kingdom, is located in what state where Cadillac Ranch can be found?
    Texas

14. Lake Borgne is a lagoon of the Gulf of Mexico in what U.S. state?
    Louisiana

15. Buffalo wings take their name from a city located on Lake Erie in which state?
    New York

16. The Mahomet Aquifer is the most important aquifer in the eastern part of what state bordering Lake Michigan to the north?
    Illinois

17. You can create your own custom PEZ candy dispenser near New Haven in which New England state?
    Connecticut

18. Mille Lacs Lake is the second largest inland lake in what U.S. state whose largest is Red Lake?
    Minnesota

19. Oreo cookies are mainly produced in Richmond in which eastern state?
    Virginia

20. The World's Tallest Thermometer is located near Death Valley and the Cronese Mountains in the city of Baker in what state?
California

21. The Shivwits Plateau is located near the Hurricane Cliffs in what state's Mohave County?
Arizona

22. Diamond Peak is the highest point in the Lemhi Mountain Range in what U.S. state?
Idaho

23. The Buttermilk Channel in southeastern New York separates what island from Brooklyn?
Governors Island

24. The Union Watersphere is near Newark Liberty International Airport in what state?
New Jersey

25. Cape Krusenstern National Monument is in northwestern Alaska along what sea?
Chukchi Sea

26. Doritos were invented in the 1960s at Disneyland in the Santa Ana Valley in which state?
California

27. The Palo Duro Canyon is located near Amarillo in what state?
    Texas

28. You can find a high concentration of hoodoos in Goblin Valley State Park in the San Rafael Swell of what state?
    Utah

29. In 1989, Hurricane Hugo tore through the southeastern United States, nearly destroying Francis Marion National Forest in what state bordering North Carolina?
    South Carolina

30. LaBarge Rock is in the shape of a column and rises above the Missouri River in Lewis and Clark National Forest in what state?
    Montana

31. The city of Omaha in Nebraska is located on what river?
    Mississippi River

32. The Dr. Pepper Museum is located in Waco near the Brazos River in which state?
    Texas

33. A Jell-O plant is located in Mason City in which state that borders the Mississippi River?
    Iowa

34. The city of Taos is located east of the Rio Grande Gorge and near the Taos Plateau Volcanic Field in what state?
    New Mexico

35. Name the deepest lake in Maine, home to Frye Island and partially in a state park of the same name.
    Sebago Lake

36. The World's Largest Buffalo Monument is a sculpture located in Jamestown, in what state bordering Montana?
    North Dakota

37. Part of the Pryor Mountains are located in Custer National Forest in what state home to Medicine Rocks State Park?
    Montana

38. Itasca Park in Minnesota is where you can find the headwaters of what major river?
    Mississippi River

39. The St. Francis River can be found flowing near Taum Sauk Mountain, the highest natural point in what state?
    Missouri

40. The Tooth of Time is a landform in the Sangre de Cristo Mountains in what state?
    New Mexico

41. The Oxnard Plain borders the Topatopa Mountains to the north and is located near what promontory in a state park of the same name?
Point Mugu

42. Potato chips are produced by the Utz factory in Hanover in which state?
Pennsylvania

43. Davis Dam forms what reservoir in the Cottonwood Valley between Arizona and Nevada?
Lake Mohave

44. The Ben & Jerry's ice cream factory is located east of Lake Champlain in which state?
Vermont

45. Lake Hartwell is situated on the border between Georgia and what other state?
South Carolina

46. Thor's Well is an odd sinkhole formation on Cape Perpetua in what northwestern state?
Oregon

47. The Pecos River originates in what state's part of the Sangre de Cristo Mountains?
New Mexico

48. Passamaquoddy Bay, located at the mouth of the St. Croix River, is an inlet of what bay that borders Maine and has the highest tidal range in the world?
    Bay of Fundy

49. Seitz Canyon can be found in the Ruby Mountains near Seitz Lake in Elko County in what state?
    Nevada

50. What state, known as the Centennial State, is home to the San Luis Valley and Glenwood Canyon?
    Colorado

51. Chauncey Peak is part of the Metacomet Ridge in what New England state?
    Connecticut

52. Kawai Nui Marsh, the largest wetland area in Hawaii, is on what island where the Lanikai Beach can be found?
    Oahu

53. The Mojave Phone Booth, a lone telephone booth that was removed in the year 2000, was located in Mojave National Preserve in what state?
    California

54. Carhenge is a replica of Stonehenge in the United Kingdom located in the High Plains Region, near the city of Alliance in what state?
    Nebraska

55. Lucy the Elephant is a 65-foot tall structure near Atlantic City in what state?
    New Jersey

56. Walpole Island is located in Canada's Ontario Province in what lake bordering Michigan to the west?
    Lake St. Clair

57. The Chamizal National Memorial is situated on the border between the United States and Mexico in what Texan city?
    El Paso

58. The Cabazon Dinosaurs are two large dinosaur sculptures near the San Gorgonio Pass in Cabazon in what state?
    California

59. Sowbelly Canyon is part of the Pine Ridge Region in what state whose panhandle is home to the Courthouse and Jail Rocks?
    Nebraska

60. You can watch underwater performances by "mermaids" at the Weeki Wachee Springs in what state bordering Alabama to the northwest?
    Florida

61. Tallulah Falls Lake is located above Tallulah Gorge in what state whose highest point is Brasstown Bald?
    Georgia

62. Czechland Lake Recreation Area and Conestoga Lake are bodies of water in what state home to the Nine Mile Prairie?
    Nebraska

63. San Timoteo Canyon is northeast of the Badlands in the San Jacinto Mountains of what state?
    California

64. The Menominee River enters Green Bay between Michigan and what other state?
    Wisconsin

65. The Balcones Fault System, which is thought to be related to the formation of the Ouachita Mountains, is located in what state?
    Texas

66. Currituck Sound, an inlet of the Atlantic Ocean, is located north of Albemarle Sound and borders North Carolina and what other state?
    Virginia

67. Pahuk Hill is situated on the Platte River in what state whose Robidoux Pass is located in the Wildcat Hills?
    Nebraska

68. The Meramec Caverns are located within the Ozark Mountains in what state?
    Missouri

69. Cavanal Hill is known as the world's tallest hill and is located in what state home to the Antelope Hills?
    Oklahoma

70. McKittrick Canyon is situated in what Texan mountain range bordering the Delaware Mountains to the south?
    Guadalupe Mountains

71. Smith Falls is located next to the Niobrara National Scenic River in what state?
    Nebraska

72. The Upper and Lower Peninsulas in Michigan are connected by what bridge that spans the Straits of Mackinac?
    Mackinac Bridge

73. Santa Rosa Sound borders the Fairpoint Peninsula in what state?
    Florida

74. The Tahquamenon Falls are a set of waterfalls near Lake Superior in what state?
    Michigan

75. Great Captain Island is situated off the coast of Greenwich and contains the southernmost point of land in New England. This island belongs to what state?
Connecticut

76. Mount Shasta is in the Siskiyou County of what state?
California

77. Lucy the Elephant is the oldest roadside tourist attraction in the United States on Absecon Island in what state?
New Jersey

78. Lake Winnipesaukee includes Paugus Bay in the Lakes Region of what state?
New Hampshire

79. Zwaanendael Colony, a Dutch colonial settlement, was built in 1631 in what state known as the "Blue Hen State"?
Delaware

80. The city of Russellville is located on Lake Dardanelle in what landlocked state?
Arkansas

81. Brasstown Bald is the highest peak in what state that is the largest state east of the Mississippi River?
Georgia

82. The Arikaree River, a tributary of the Republican River, has its source in Elbert County in what state?
Colorado

83. The Bayou Corne Sinkhole is located in the Assumption Parish of what state?
Louisiana

84. What state is known as "America's Dairyland", and has Timms Hill as its highest point?
Wisconsin

85. Waugoshance Point borders Sturgeon Bay on the northwestern coast of what peninsula in Michigan?
Lower Peninsula

86. What Hawaiian island is famously regarded as "The Gathering Place" and is home to the Hawaii Cryptologic Center and Kualoa Regional Park?
Oahu

87. Mono Lake is a saline soda lake in the Mono Basin of what state home to the Santa Cruz Mountains?
California

88. What state preserves Puebloan structures at Aztec Ruins National Monument near the city of Aztec?
New Mexico

89. Butterfield Canyon is located in the Oquirrh Mountains of what state home to Delicate Arch and Kings Peak?
Utah

90. Turner Falls is located in the Arbuckle Mountains, an ancient mountain range in what state?
Oklahoma

91. The Monument Rocks, also known as the Chalk Pyramids, are located in what state home to Big Basin Prairie Reserve?
Kansas

92. Attu Island, the largest of the Near Island Group, is home to a cape that is the westernmost point in the United States. Name this cape.
Cape Wrangell

93. The Brewster Flats are located on what bay in Massachusetts near the Race Point Lighthouse?
Cape Cod Bay

94. The Courthouse and Jail Rocks are in the North Platte River Valley in what Midwestern state?
Nebraska

95. A dam separates Paugus Bay and Opechee Bay in the Lakes Region of what northeastern state bordering Canada?
New Hampshire

96. Pioneer Courthouse Square is located in the downtown area of Portland in what state?
    Oregon

97. Augusta, Georgia, is situated on what river formed by the confluence of the Seneca and Tugaloo Rivers?
    Savannah River

98. The town of Simmesport is located at the confluence of the Atchafalaya and Red Rivers in what state?
    Louisiana

99. The Rio Grande carves out the Santa Elena Canyon in what Texan national park?
    Big Bend National Park

100. The Pymatuning Reservoir is located on the border between Pennsylvania and what state to its west?
    Ohio

United States State Fact Files

ALABAMA

Capital: Montgomery

Major Cities: Birmingham, Montgomery, Mobile, Huntsville, Tuscaloosa, Dothan, Auburn

Population: 4,870,000

Area: 52,419 sq mi

Bordering States: Florida, Georgia, Mississippi, Tennessee

Nicknames: Yellowhammer State, Cotton State, Heart of Dixie

Mountain Ranges: Appalachian Mountains, Florida Ridge Hills

Rivers: Tennessee River, Chattahoochee River, Alabama River, Coosa River, Tallapoosa River, Tombigbee River, Conecuh River, Elk River, Cahaba River, Black Warrior River

Lakes and Reservoirs: Bankhead Lake, Lake Martin, Weiss Lake, Wheeler Lake

Bays and Gulfs: Mobile Bay, Gulf of Mexico, Perdido Bay

Straits: Perdido Pass

Plateaus: Cumberland Plateau, Piedmont Plateau, Sand Mountain Plateau Region

Peninsulas, Capes, and Points: Mobile Point

Islands: Dauphin Island, Gaillard Island, Ono Island

Caves: DeSoto Caverns

Canyons and Gorges: Dismals Canyon, Tennessee River Gorge

Swamps, Marshes, and Wetlands: Beaver Creek Swamp, Mobile-Tensaw River Delta

Valleys: Great Appalachian Valley, Tennessee Valley, Sequatchie Valley

Other Landforms: Lookout Mountain, Red Mountain

Other Bodies of Water: Mississippi Sound, Tennessee-Tombigbee Waterway, Intracoastal Waterway

Mountains: Cheaha Mountain, Monte Sano Mountain, Red Mountain

National Forests: Conecuh National Forest, Talladega National Forest, Tuskegee National Forest, William B. Bankhead National Forest

National Monuments: Birmingham Civil Rights National Monument, Freedom Rights National Monument, Russell Cave National Monument

Other Sites: USS Alabama, U.S. Space and Rocket Center, Birmingham Civil Rights Institute, Vulcan Statue, Point Mallard Park, Sloss Furnaces, Huntsville Botanical Garden, Gulf State Park, Fort Gaines, Cheaha State Park, Noccalula Falls Park, Ruffner Mountain Nature Preserve

International/Major Airports: Birmingham-Shuttlesworth International Airport, Huntsville International Airport

ALASKA

Capital: Juneau

Major Cities: Anchorage, Juneau, Fairbanks, Sitka, Ketchikan

Population: 740,000

Area: 665,384 sq mi

Nicknames: The Last Frontier

Mountain Ranges: Alaska Range, Brooks Range, Aleutian Range, Coast Mountains, St. Elias Mountains, Chugach Mountains, Wrangell Mountains, Kenai Mountains, Kuskokwim Mountains, Chilkat Range

Rivers: Yukon River, Kuskokwim River, Porcupine River, Tanana River, Innoko River, Koyukuk River

Lakes and Reservoirs: Kenai Lake, Iliamna Lake, Skilak Lake, Twin Lakes, Tangle Lakes, Lake Clark, Becharof Lake

Bays and Gulfs: Gulf of Alaska, Glacier Bay, Yakutat Bay, Disenchantment Bay, Bristol Bay, Nushagak Bay, Kachemak Bay

Seas: Bering Sea, Beaufort Sea

Straits: Dixon Entrance, Peril Strait, Shelikof Strait, Chatham Strait, Ice Strait, Stephens Passage, Sumner Strait

Plateaus: York Plateau

Peninsulas, Capes, and Points: Alaska Peninsula, Seward Peninsula, Kenai Peninsula, Cleveland Peninsula, Lindenberg Peninsula, Glass Peninsula, Cape Wrangell, Cape Krusenstern

Isthmuses and Spits: Homer Spit

Islands: Kodiak Island, Prince of Wales Island, Revillagigedo Island, Chichagof Island, Baranof Island, Admiralty Island, Annette Island, Kuiu Island, Kupreanof Island, Mitkof Island, Woewodski Island, Etolin Island

Archipelagoes: Aleutian Islands, Alexander Archipelago

Canyons and Gorges: Keystone Canyon

Valleys: Tanana Valley, Mendenhall Valley

Other Landforms: Malaspina Glacier, Bering Glacier, Hubbard Glacier, Nabesna Glacier, Bagley Icefield, Russell Fjord

Other Bodies of Water: Cook Inlet, Kotzebue Sound, Prince William Sound, Frederick Sound, Moira Sound, Lynn Canal, Duncan Canal, Seymour Canal, Turnagain Arm, Knik Arm, Turnagain Arm

Mountains: Denali, Mount St. Elias, Mount Foraker, Mount Bona, Mount Blackburn, Mount Sanford, Mount Vancouver, Mount Fairweather, Mount Hubbard, Mount Bear, Mount Shishaldin

National Parks: Denali National Park, Gates of the Arctic National Park, Glacier Bay National Park, Katmai National Park, Kenai Fjords National Park, Kobuk Valley National Park, Lake Clark National Park, Wrangell-St. Elias National Park

National Forests: Chugach National Forest, Tongass National Forest

National Monuments: Admiralty Island National Monument, Aniakchak National Monument, Cape Krusenstern National Monument, Misty Fjords National Monument, Word War II Valor in the Pacific National Monument

Other Sites: Prudhoe Bay Oil Field, Alaska Native Heritage Center, Point Woronzof Park, Chugach State Park

International/Major Airports: Ted Stevens Anchorage International Airport, Fairbanks International Airport, Juneau International Airport

ARIZONA

Capital: Phoenix

Major Cities: Phoenix, Tucson, Mesa, Chandler, Gilbert, Glendale, Scottsdale, Tempe, Peoria, Surprise

Population: 6,940,000

Area: 113,990 sq mi

Bordering States: California, Utah, Nevada, New Mexico, Colorado

Nicknames: Grand Canyon State, Apache State, Aztec State

Mountain Ranges: Santa Catalina Mountains, Santa Rita Mountains, Mazatzal Mountains, McDowell Mountains, Rucson Mountains, Rincon Mountains, Chiricahua Mountains, Superstition Mountains, San Francisco Peaks, White Mountains, Hualapai Mountains, Dome Rock Mountains

Rivers: Colorado River, Gila River, Little Colorado River, Salt River, Santa Cruz River, Verde River

Lakes and Reservoirs: Lake Powell, Lake Mead, Lake Havasu, Lake Mohave, Theodore Roosevelt Lake, San Carlos Lake

Deserts: Sonoran Desert, Painted Desert, Yuma Desert, Tonopah Desert, Lechuguilla Desert, Mojave Desert, Chihuahuan Desert

Plateaus: Colorado Plateau, Mogollon Plateau, Kaibab Plateau

Plains and Basins: Cactus Plain

Caves: Grand Canyon Caverns, Kartchner Caverns

Canyons and Gorges: Grand Canyon, Antelope Canyon, Glen Canyon, Oak Creek Canyon, Sabino Canyon, Marble Canyon

Valleys: Monument Valley, Mohave Valley, Chinle Valley, Cienega Valley, Parker Valley

Swamps, Marshes, and Wetlands: Tres Rios Wetlands

Other Landforms: Vermilion Cliffs, The Wave, Chinle Formation, Moenkopi Formation, Hurricane Cliffs, Tonto Natural Bridge

Other Bodies of Water: Havasu Falls, Havasu Creek, Grand Falls

Mountains: Humphreys Peak, Agassiz Peak, Fremont Peak, Mount Baldy, Doyle Peak, Camelback Mountain, Mount Lemmon

National Parks: Grand Canyon National Park, Petrified Forest National Park, Saguaro National Park

National Forests: Apache-Sitgreaves National Forest, Coconino National Forest, Coronado National Forest, Kaibab National Forest, Prescott National Forest, Tonto National Forest

National Monuments: Agua Fria National Monument, Canyon de Chelly National Monument, Casa Grande Ruins National Monument, Chiricahua National Monument, Grand Canyon-Parashant National Monument, Hohokam Pima National Monument, Ironwood Forest National Monument, Montezuma Castle National Monument, Navajo National Monument, Organ Pipe Cactus National Monument, Pipe Spring National Monument, Sonoran Desert National Monument, Sunset Crater Volcano National Monument, Tonto National Monument, Tuzigoot National Monument, Vermilion Cliffs National Monument, Walnut Canyon National Monument, Wupatki National Monument

Other Sites: Meteor Crater National Natural Landmark, Biosphere 2, Desert Botanical Garden, Lowell Observatory, Four Corners Monument

Dams: Hoover Dam, Glen Canyon Dam, New Weddell Dam, Theodore Roosevelt Dam, Bartlett Dam, Parker Dam

International/Major Airports: Phoenix Sky Harbor International Airport, Tucson International Airport, Phoenix-Mesa Gateway Airport

ARKANSAS

Capital: Little Rock

Major Cities: Little Rock, Fort Smith, Fayetteville, Springdale, Jonesboro

Population: 2,990,000

Area: 53,178 sq mi

Bordering States: Texas, Louisiana, Mississippi, Oklahoma, Missouri, Tennessee

Nicknames: Natural State

Mountain Ranges: Ozark Mountains, Ouachita Mountains, Boston Mountains, U.S. Interior Highlands

Rivers: Mississippi River, Arkansas River, Red River of the South, White River, Ouachita River, St. Francis River, Black River, Cache River, Saline River, Little Missouri River

Lakes and Reservoirs: Bull Shoals Lake, Table Rock Lake, Greers Ferry Lake, Lake Dardanelle, Lake Ouachita, Millwood Lake, Beaver Lake, DeGray Lake, Lake Conway

Plateaus: Ozark Plateau

Plains and Basins: Gulf Coastal Plain, Arkansas Delta, Mississippi Alluvial Plain

Caves: Blanchard Springs Caverns

Valleys: Arkansas River Valley

Other Landforms: Crowley's Ridge, Piney Woods

Other Bodies of Water: Bayou Bartholomew

Mountains: Mount Magazine, Mount Nebo, Petit Jean Mountain, Shinall Mountain

National Parks: Hot Springs National Park

National Forests: Ouachita National Forest, Ozark National Forest, St. Francis National Forest

Other Sites: Black Fork Mountain Wilderness, Bathhouse Row, Pea Ridge National Military Park, Arkansas Post National Memorial, Crater of Diamonds State Park, Pinnacle Mountain State Park, Buffalo National River, Toltec Mounds Archaeological State Park, War Memorial Stadium, Missouri and Northern Arkansas Railroad

Dams: Bull Shoals Dam

International/Major Airports: Bill and Hillary Clinton National Airport

CALIFORNIA

Capital: Sacramento

Major Cities: Los Angeles, San Diego, San Jose, San Francisco, Fresno, Sacramento, Long Beach, Oakland, Bakersfield, Anaheim, Santa Ana, Riverside, Stockton, Chula Vista, Irvine, Fremont, San Bernardino, Modesto, Fontana, Oxnard, Moreno Valley, Huntington Beach, Glendale

Population: 39,300,000

Area: 163,694 sq mi

Bordering States: Oregon, Nevada, Arizona

Nicknames: The Golden State, El Dorado State

Mountain Ranges: Sierra Nevada, Cascade Range, Klamath Mountains, San Gabriel Mountains, Santa Cruz Mountains, San Bernardino Mountains, Laguna Mountains, San Jacinto Mountains, Santa Rosa Mountains, Santa Ana Mountains, Santa Susana Mountains, Tehachapi Mountains, Santa Monica Mountains, Diablo Mountains, Chocolate Mountains, Orocopia Mountains, Chuckwalla Mountains, White Mountains, Inyo Mountains, Trinity Alps, Santa Ynez Mountains, Topatopa Mountains, Sierra Madre, Sierra Pelona, Chemehuevi Mountains, Whipple Mountains, Providence Mountains

Rivers: Colorado River, Sacramento River, San Joaquin River, Klamath River, Pit River, Eel River, Amargosa River, Salinas River, American River

Lakes and Reservoirs: Lake Tahoe, Salton Sea, Owens Lake, Mono Lake, Lexington Reservoir, Clear Lake, Shasta Lake, Trinity Lake

Bays and Gulfs: Gulf of Santa Catalina, San Francisco Bay, Monterey Bay, Gulf of the Farallones, Suisun Bay, Humboldt Bay, Richardson Bay, San Diego Bay, Santa Monica Bay

Straits: Carquinez Strait, Oakland Estuary, Santa Barbara Channel

Deserts: Mojave Desert, Sonoran Desert, Colorado Desert, Yuha Desert, Amargosa Desert

Plateaus: Modoc Plateau, Santa Rosa Plateau

Plains and Basins: Oxnard Plain, Great Basin

Peninsulas, Capes, and Points: Monterey Peninsula, Balboa Peninsula, San Francisco Peninsula, Palo Verdes Peninsula, Tiburon Peninsula, Marin Headlands, Point Reyes, Point Dume, Point Loma

Isthmuses and Spits: Silver Strand, Isthmus of Santa Catalina

Islands: Santa Catalina Island, Santa Barbara Island, Santa Cruz Island, Santa Rosa Island, San Miguel Island, San Nicolas Island, San Clemente Island, Balboa Island, Terminal Island, Anacapa Island

Archipelagoes: Channel Islands, Farallon Islands

Caves: Lake Shasta Caverns, Mitchell Caverns, Black Chasm Cave

Canyons and Gorges: Laguna Canyon, San Timoteo Canyon, San Clemente Canyon, Black Star Canyon, Soledad Canyon

Swamps, Marshes, and Wetlands: Suisun Marsh, Madrona Marsh, Goleta Slough, Los Cerritos Wetlands, Ballona Wetlands

Valleys: Central Valley, San Joaquin Valley, Yosemite Valley, Death Valley, Owens Valley, Antelope Valley, Santa Clara River Valley, San Fernando Valley, Santa Clarita Valley, Simi Valley, Coachella Valley, Imperial Valley, Saline Valley, Eureka Valley

Other Landforms: Moenkopi Formation, Redonda Mesa, Goat Rock Beach, Natural Bridges State Beach, North Dome, Half Dome, Glacier Point, Cajon Pass, San Francisquito Pass, Sutter Buttes, Red Cones, Salton Buttes, Mono-Inyo Craters, Pisgah Crater, Algodones Dunes, Kelso Dunes, Whitney Glacier

Other Bodies of Water: Santa Clara Valley Aquifer, All-American Canal

Mountains: Mount Whitney, Mount Shasta, North Palisade, White Mountain Peak, Mount Williamson, Mammoth Mountain, San Jacinto Peak, Lassen Peak, Mount Humphreys, Mount San Antonio, Mount Langley, Glass Mountain, Mount Tom, Mount Lyell, Red Slate Mountain

National Parks: Sequoia National Park, Redwood National Park, Klamath National Park, Death Valley National Park, Joshua Tree National Park, Pinnacles National Park, Channel Islands National Park, Lassen Volcanic National Park

National Forests: Angeles National Forest, Cleveland National Forest, Eldorado National Forest, Humboldt-Toiyabe National Forest, Inyo National Forest, Klamath National Forest, Lassen National Forest, Los Padres National Forest, Mendocino National Forest, Modoc National Forest, Plumas National Forest, Rogue River-Siskiyou National Forest, San Bernardino National Forest, Sequoia National Forest, Shasta-Trinity National Forest, Sierra National Forest, Six Rivers National Forest, Stanislaus National Forest, Tahoe National Forest

National Monuments: Berryessa Snow Mountain National Monument, Cabrillo National Monument, California Coastal National Monument, Carrizo Plain National Monument, Castle Mountains National Monument, Cesar E. Chavez National Monument, Devils Postpile National Monument, Fort Ord National Monument, Giant Sequoia National Monument, Lava Beds National Monument, Mojave Trails National Monument, Muir Woods National Monument, San Gabriel Mountains National Monument, Sand to Snow National Monument, Santa Rosa and San Jacinto Mountains National Monument, World War II Valor in the Pacific National Monument

Other Sites: Golden Gate Bridge, Disneyland, Hollywood, Alcatraz Island, Balboa Park, Legoland California, SeaWorld San Diego, Griffith Observatory, Hollywood Boulevard, San Diego Zoo, Monterey Bay Aquarium, Hearst Castle, La Brea Tar Pits, Santa Monica Pier, USS Midway Museum, Lombard Street, Bay Bridge, Union Square, Squaw Valley Sea Resort

Dams: Oroville Dam, Shasta Dam, New Melones Dam, Pine Flat Dam, Trinity Dam, New Bullards Bar Dam, New Don Pedro Dam, Seven Oaks Dam, New Exchequer Dam

International/Major Airports: Los Angeles International Airport, San Francisco International Airport, San Diego International Airport, Oakland International Airport, John Wayne Airport, Norman Y. Mineta San Jose International Airport, Sacramento International Airport, Ontario International Airport, Bob Hope Airport, Long Beach Airport

COLORADO

Capital: Denver

Major Cities: Denver, Colorado Springs, Aurora, Fort Collins, Lakewood, Thornton, Arvada, Westminster

Population: 5,550,000

Area: 104,093 sq mi

Bordering States: Utah, Arizona, New Mexico, Wyoming, Oklahoma, Kansas, Nebraska

Nicknames: Centennial State Highest State

Mountain Ranges: Rocky Mountains, Sawatch Range, Sangre de Cristo Mountains, San Juan Mountains, Mosquito Range, Front Range, Elk Mountains, Park Rage, Medicine Bow Mountains, Culebra Range, Gore Range

Rivers: Rio Grande, Arkansas River, Colorado River, Canadian River, Green River, North Platte River, Cimarron River, South Platte River

Lakes and Reservoirs: Blue Mesa Reservoir, John Martin Reservoir, Lake Granby

Plateaus: Colorado Plateau, Uncompahgre Plateau

Caves: Cave of the Winds

Canyons and Gorges: Glenwood Canyon, Rattlesnake Canyon, Royal Gorge

Swamps, Marshes, and Wetlands: Blanca Wetlands

Valleys: San Luis Valley

Other Landforms: Moenkopi Formation, Collegiate Peaks, Colorado Mineral Belt

Other Bodies of Water: Bridal Veil Falls

Mountains: Mount Elbert, Mount Evans, Pikes Peak, Longs Peak

National Parks: Black Canyon of the Gunnison National Park, Great Sand Dunes National Park, Mesa Verde National Park, Rocky Mountain National Park

National Forests: Arapaho National Forest, Grand Mesa National Forest, Gunnison National Forest, Pike National Forest, Rio Grande National Forest, Roosevelt National Forest, Routt National Forest, San Isabel National Forest, San Juan National Forest, Uncompahgre National Forest, White River National Forest

National Monuments: Browns Canyon National Monument, Canyons of the Ancients National Monument, Chimney Rock National Monument, Colorado National Monument, Dinosaur National Monument, Florissant Fossil Beds National Monument, Hovenweep National Monument, Yucca House National Monument

Other Sites: Cliff Palace, Manitou Cliff Dwellings, Chapin Mesa Archaeological Museum, Front Range Urban Corridor, Comanche National Grassland, Cimarron National Grassland, North Cheyenne Canon Park, San Juan Volcanic Field

Dams: Blue Mesa Dam

International/Major Airports: Denver International Airport, City of Colorado Springs Municipal Airport

CONNECTICUT

Capital: Hartford

Major Cities: Bridgeport, New Haven, Stamford, Hartford, Waterbury

Population: 3,580,000

Area: 5,543 sq mi

Bordering States: New York, Massachusetts, Rhode Island

Nicknames: The Constitution State, The Nutmeg State, The Provisions State

Mountain Ranges: Taconic Mountains, Berkshire Hills, Metacomet Ridge, Hanging Hills

Rivers: Connecticut River, Housatonic River, Quinnebaug River, Farmington River, Quinnipiac River

Lakes and Reservoirs: Candlewood Lake, Lake Lillinonah, Twin Lakes, Mashapaug Lake, Pinewood Lake

Bays and Gulfs: Little Narragansett Bay

Peninsulas, Capes, and Points: Shippan Point, City Point

Islands: Mason's Island, Calf Island

Archipelagoes: Thimble Islands, Norwalk Islands

Swamps, Marshes, and Wetlands: Beckley Bog

Valleys: Connecticut River Valley, Housatonic Valley

Other Landforms: West Rock Ridge, Sleeping Giant, Coney Rock, Peter's Rock, Totoket Mountain

Other Bodies of Water: Long Island Sound, Intracoastal Waterway, Wethersfield Cove, New Haven Harbor, Farmington Canal, Enfield Falls Canal

Mountains: Mount Frissell, Bear Mountain, Saltonstall Mountain, Chauncey Peak, Lamentation Mountain

Other Sites: Mystic Seaport, Stewart B. McKinney National Wildlife Refuge, Dinosaur State Park, Gillette Castle, Mark Twain House, Hammonasset Beach State Park, Elizabeth Park, USS Nautilus, Amistad Memorial

Dams: Saville Dam

International/Major Airports: Bradley International Airport, Tweed New Haven Airport

DELAWARE

Capital: Dover

Major Cities: Wilmington, Dover, Newark, Middletown, Smyrna

Population: 950,000

Area: 2,488 sq mi

Bordering States: Maryland, Pennsylvania, New Jersey

Nicknames: The First State, Blue Hen State

Rivers: Delaware River, Choptank River, Pocomoke River, Nanticoke River

Lakes and Reservoirs: Newark Reservoir, Hoopes Reservoir

Bays and Gulfs: Delaware Bay, Rehoboth Bay

Plateaus: Piedmont Plateau

Plains and Basins: Atlantic Coastal Plain

Peninsulas, Capes, and Points: Delmarva Peninsula, Cape Henlopen

Isthmuses and Spits: Fenwick Island

Caves: Beaver Valley Rock Shelter Site

Swamps, Marshes, and Wetlands: Great Cypress Swamp

Other Bodies of Water: Chesapeake and Delaware Canal

Other Sites: Ebright Azimuth, Twelve-Mile Circle, Bombay Hook National Wildlife Refuge, Prime Hook National Wildlife Refuge, First State Heritage Park

FLORIDA

Capital: Tallahassee

Major Cities: Jacksonville, Miami, Tampa, Orlando, St. Petersburg, Hialeah, Fort Lauderdale, Tallahassee, Port St. Lucie, Cape Coral, Pembroke Pines, Hollywood

Population: 20,620,000

Area: 65,757 sq mi

Bordering States: Alabama, Georgia

Nicknames: Sunshine State, Peninsula State, Orange State, Gulf State, Alligator State

Rivers: Chattahoochee (Apalachicola) River, St. Johns River, Suwannee River, Ochlockonee River, Alapaha River

Lakes and Reservoirs: Lake Okeechobee, Lake Seminole, Lake Kissimmee, Lake Apopka, Lake Tohopekaliga

Bays and Gulfs: Gulf of Mexico, Tampa Bay, Biscayne Bay, Boca Ciega Bay, Florida Bay, Apalachicola Bay, Pensacola Bay, St. Joseph Bay, Charlotte Harbor, Choctawhatchee Bay, St. Andrews Bay, Escambia Bay, East Bay, Whitewater Bay, Ponce de Leon Bay

Straits: Straits of Florida

Plains and Basins: Woodville Karst Plain, Okeechobee Plain, DeSoto Plain

Peninsulas, Capes, and Points: Florida Peninsula, Cape Canaveral, Fairpoint Peninsula, Pinellas Peninsula, Cape Sable, Cape Romano, St. Joseph Peninsula, Cape San Blas

Islands: Key West, Merritt Island, Pine Island, Bahia Honda Key, Santa Rosa Island, Key Largo, Sanibel Island, Captiva Island, St. Vincent Island, St. George Island, Dog Island

Archipelagoes: Florida Keys, Dry Tortugas, Marquesas Keys, Ten Thousand Islands, Sea Islands

Swamps, Marshes, and Wetlands: Everglades, Okefenokee Swamp, Wakodahatchee Wetlands

Other Landforms: Britton Hill, Sugarloaf Mountain, Leon Sinks Geological Area

Other Bodies of Water: Gulf Intracoastal Waterway, Indian River Lagoon, Mosquito Lagoon, Santa Rosa Sound, Pine Island Sound, Matanzas Inlet, Wakulla Springs, Weeki Wachee Springs

National Parks: Everglades National Park, Dry Tortugas National Park, Biscayne National Park

National Forests: Apalachicola National Forest, Ocala National Forest, Osceola National Forest

National Monuments: Castillo de San Marcos National Monument, Fort Matanzas National Monument

Other Sites: Florida Caverns State Park, Seven Mile Bridge, Kennedy Space Center, Canaveral National Seashore, Universal Studios Florida, Walt Disney World, Daytona International Speedway, Brickell Financial District, Flagler Memorial Island, SS American Victory, Jacksonville Skyway, Villa Vizcaya, Freedom Tower, Memorial Park, Amalie Arena

Dams: Franklin Lock and Dam

International/Major Airports: Orlando International Airport, Miami International Airport, Tampa International Airport, Fort Lauderdale-Hollywood International Airport, Southwest Florida International Airport, Palm Beach International Airport, Jacksonville International Airport

GEORGIA

Capital: Atlanta

Major Cities: Atlanta, Columbus, Augusta, Macon, Savannah

Population: 10,315,000

Area: 59,425 sq mi

Bordering States: Alabama, Florida, South Carolina, Tennessee, North Carolina

Nicknames: Peach State, Empire State of the South, Goober State

Mountain Ranges: Appalachian Mountains, Blue Ridge Mountains, Cohutta Mountains

Rivers: Chattahoochee River, Flint River, Savannah River, Ogeechee River, Ocmulgee River, Coosa River

Lakes and Reservoirs: Lake Strom Thurmond, Lake Burton, Lake Hartwell, Walter F. George Lake, Lake Seminole, Lake Oconee

Plateaus: Cumberland Plateau

Islands: Cumberland Island

Archipelagoes: Golden Isles, Sea Islands

Caves: Petty John's Cave

Canyons and Gorges: Tallulah Gorge

Swamps, Marshes, and Wetlands: Okefenokee Swamp

Valleys: Great Appalachian Valley

Other Bodies of Water: Intracoastal Waterway

Mountains: Cold Mountain

National Forests: Chattahoochee-Oconee National Forest

National Monuments: Fort Frederica National Monument, Fort Pulaski National Monument, Ocmulgee National Monument

Other Sites: Coca-Cola Headquarters, Centennial Olympic Park, Millennium Gate Museum, Martin Luther King Jr. National Historic Site, Rock City, Atlanta Botanical Garden, Millennium Gate Museum, The Phoenix Statue and Monument, Georgia Aquarium

Dams: Carters Dam

International/Major Airports: Hartsfield-Jackson Atlanta International Airport, Savannah/Hilton Head International Airport

HAWAII

Capital: Honolulu

Major Cities: Honolulu, Hilo, Kailua, Kapolei, Kaneohe

Population: 1,430,000

Area: 10,931 sq mi

Nicknames: Aloha State, Paradise of the Pacific

Mountain Ranges: Koolau Range

Rivers: Wailuku River, Kaukonahua River, Hanalei River

Lakes and Reservoirs: Lake Waiau, Ka Loko Reservoir

Bays and Gulfs: Kane'ohe Bay, Hanauma Bay, Hanalei Bay, Kealakekua Bay

Straits: 'Au'au Channel, 'Alenuihaha Channel, Kealaikahiki Channel

Peninsulas, Capes, and Points: Kalaupapa Peninsula, Magic Island, Mokapu Peninsula

Islands: Big Island (Hawaii), Oahu, Maui, Kauai, Lanai, Molokai, Niihau, Kahoolawe, Nihoa, Molokini, Lisianski Island, Lihua, Necker Island, Kure Atoll, Ka'ula, Tern Island, Moku Manu, Coconut Island

Archipelagoes: Hawaiian Islands

Caves: Makauwahi Cave

Canyons and Gorges: Waimea Canyon

Swamps, Marshes, and Wetlands: Kawainui Marsh

Valleys: Iao Valley, Waipio Valley, Pololu Valley, Waimea Valley

Other Landforms: Diamond Head, Gardner Pinnacles, Punalu'u Beach, Nu'uanu Pali

Mountains: Mauna Kea, Mauna Loa, Haleakala, Kohala Mountain, Hualalai, Pu'u Kukui, Mauna Kahalawai (West Maui Volcano), Kawaikini, Mount Waialeale

National Parks: Hawaii Volcanoes National Park, Haleakala National Park

National Monuments: Honouliuli National Monument, Papahanaumokuakea Marine National Monument, World War II Valor in the Pacific National Monument

Other Sites: Ala Moana Beach Park, Ala Moana Center, Kualoa Regional Park, Na Pali Coast State Park, Hawaii Volcano Observatory, National Memorial Cemetery of the Pacific, Pu'u Loa Petroglyphs, Pu'uhonua o Honaunau National Historical Park, Pu'ukohola Heiau National Historic Site, Keahiakawelo (Garden of the Gods)

International/Major Airports: Honolulu International Airport, Kahului Airport, Kona International Airport, Lihue Airport

IDAHO

Capital: Boise

Major Cities: Boise, Nampa, Meridian, Idaho Falls, Pocatello

Population: 1,680,000

Area: 83,568 sq mi

Bordering States: Washington, Oregon, Wyoming, Montana, Nevada, Utah

Nicknames: Gem State

Mountain Ranges: Rocky Mountains, Bitterroot Range, Sawtooth Range, Columbia Mountains, Pioneer Mountains, Lemhi Range, Selkirk Mountains, Salmon River Mountains, Smoky Mountains, Boulder Mountains, Lost River Range, Seven Devils Mountains, Soldier Mountains

Rivers: Snake River, Bear River, Kootenai River, Salmon River, Owyhee River

Lakes and Reservoirs: Lake Pend Oreille, Bear Lake, Lake Coeur d'Alene, Priest Lake

Deserts: Owyhee Desert

Plateaus: Columbia Plateau

Plains and Basins: Snake River Plain, Columbia Basin, Hell's Half Acre Lava Field

Caves: Minnetonka Cave, Wilson Butte Cave

Canyons and Gorges: Hells Canyon, Snake River Canyon

Swamps, Marshes, and Wetlands:

Valleys: Teton Valley, Sawtooth Valley, Wood River Valley, Cache Valley

Other Landforms: Menan Buttes, Henry's Fork Caldera, Island Park Caldera, Big Southern Butte, Galena Summit

Other Bodies of Water: Shoshone Falls

Mountains: Borah Peak, Leatherman Peak, Diamond Peak, Mount Idaho, Hyndman Peak, Mount Corruption, Castle Peak, Ryan Peak

National Parks: Yellowstone National Park

National Forests: Bitterroot National Forest, Boise National Forest, Caribou-Targhee National Forest, Clearwater National Forest, Idaho Panhandle National Forest, Kootenai National Forest, Nez Perce National Forest, Payette National Forest, Salmon-Challis National Forest, Sawtooth National Forest, Uinta-Wasatch-Cache National Forest, Wallowa-Whitman National Forest

National Monuments: Craters of the Moon National Monument, Hagerman Fossil Beds National Monument

Other Sites: Minidoka National Historic Site, Perrine Bridge, City of Rocks National Reserve, World Center for Birds of Prey

Dams: Dworshak Dam, Arrowrock Dam, Hells Canyon Dam, Teton Dam

International/Major Airports: Boise Airport

ILLINOIS

Capital: Springfield

Major Cities: Chicago, Aurora, Rockford, Joliet, Naperville, Springfield

Population: 12,800,000

Area: 57,913 sq mi

Bordering States: Indiana, Missouri, Kentucky, Michigan, Wisconsin, Iowa

Nicknames: Prairie State, Land of Lincoln

Mountain Ranges: Shawnee Hills

Rivers: Ohio River, Wabash River, Kaskaskia River, Mississippi River, Rock River, Illinois River, Sangamon River, Little Wash River

Lakes and Reservoirs: Lake Michigan, Carlyle Lake

Caves: Illinois Caverns

Valleys: Ohio River Valley

Other Landforms: Niagara Escarpment, Charles Mound

National Forests: Shawnee National Forest

National Monuments: Pullman National Monument

Other Sites: Garden of the Gods Wilderness, Illinois Caverns State Natural Area, Abraham Lincoln Presidential Library and Museum, Lincoln Home National Historic Site, Soldier Field, James R. Thompson Center

International/Major Airports: Chicago O'Hare International Airport, Chicago Midway International Airport

KANSAS

Capital: Topeka

Major Cities: Wichita, Overland Park, Kansas City, Olathe, Topeka

Population: 2,910,000

Area: 82,278 sq mi

Bordering States: Colorado, Oklahoma, Missouri, Nebraska

Nicknames: Sunflower State

Mountain Ranges: Ozark Mountains

Rivers: Arkansas River, Missouri River, Cimarron River, Smoky Hill River, Republican River, Neosho River, Saline River

Lakes and Reservoirs: Milford Lake, Tuttle Creek Lake, Waconda Lake, Cedar Bluff Reservoir

Plains and Basins: Great Plains

Swamps, Marshes, and Wetlands: Cheyenne Bottoms, Haskell-Baker Wetlands

Other Landforms: Flint Hills, Monument Rocks (Chalk Pyramids), Castle Rock, Mount Sunflower, Red Hills

Other Sites: Tallgrass Prairie National Preserve, Mushroom Rock State Park, Big Basin Prairie Reserve, Kanopolis State Park

Dams: Cedar Bluff Dam

International/Major Airports: Wichita Dwight D. Eisenhower National Airport

KENTUCKY

Capital: Frankfurt

Major Cities: Louisville, Lexington, Bowling Green, Owensboro, Covington

Population: 4,440,000

Area: 40,407 sq mi

Bordering States: Ohio, Tennessee, Indiana, Illinois, Virginia, Missouri, West Virginia

Nicknames: Bluegrass State

Mountain Ranges: Appalachian Mountains, Cumberland Mountains

Rivers: Ohio River, Mississippi River, Cumberland River, Tennessee River, Green River, Kentucky River

Lakes and Reservoirs: Kentucky Lake, Lake Cumberland, Lake Barkley

Plateaus: Cumberland Plateau, Pennyroyal Plateau, Appalachian Plateau

Canyons and Gorges: Red River Gorge, Kentucky River Palisades

Other Bodies of Water: Cumberland Falls

Mountains: Pine Mountain

National Parks: Mammoth Cave National Park

National Forests: Daniel Boone National Forest

International/Major Airports: Cincinnati/Northern Kentucky International Airport, Louisville International Airport

LOUISIANA

Capital: Baton Rouge

Major Cities: New Orleans, Baton Rouge, Shreveport, Lafayette, Lake Charles

Population: 4,690,000

Area: 52,378 sq mi

Bordering States: Texas, Arkansas, Mississippi

Nicknames: Pelican State, Bayou State, Sportsman's Paradise

Rivers: Mississippi River, Red River, Ouachita River, Sabine River, Pearl River

Lakes and Reservoirs: Lake Pontchartrain, Toledo Bend Reservoir, Lake Claiborne, Lake Maurepas, Caddo Lake, Lake D'Arbonne, Calcasieu Lake, Catahoula Lake, Lake St. Catherine

Bays and Gulfs: Gulf of Mexico, Atchafalaya Bay

Straits: Rigolets Strait, Chef Menteur Pass

Plains and Basins: Gulf Coastal Plain, Mississippi Alluvial Plain

Islands: Marsh Island

Archipelagoes: Chandeleur Islands

Swamps, Marshes, and Wetlands: Atchafalaya Swamp

Other Bodies of Water: Lake Borgne (lagoon), Breton Sound, Chief Menteur Pass, Rigolets Strait

Mountains: Driskill Mountain, Bayou Corne Sinkhole

National Forests: Kisatchie National Forest

National Monuments: Poverty Point National Monument

Other Sites: French Quarter, National WWII Museum, Mardi Gras World, D'Arbonne National Wildlife Refuge, Cameron Prairie National Wildlife Refuge, Jean Lafitte National Historical Park, Indian Creek Recreation Area, Bayou Sauvage National Wildlife Refuge, Mercedes-Benz Superdome, Jackson Square, Royal Street, St. Louis Cathedral, Mid-City

Dams: Sibley Lake Dam

International/Major Airports: Louis Armstrong New Orleans International Airport

MAINE

Capital: Augusta

Major Cities: Portland, Lewiston, Bangor, South Portland, Auburn

Population: 1,331,000

Area: 35,379 sq mi

Bordering States: New Hampshire

Nicknames: Pine Tree State, Vacationland

Mountain Ranges: Longfellow Mountains, Notre Dame Mountains, Mahoosuc Range, Baldface-Royce Range, Western Maine Mountains, White Mountains

Rivers: Saint John River, Androscoggin River, Kennebec River, Saco River, Penobscot River

Lakes and Reservoirs: Moosehead Lake, Sebago Lake, Chesuncook Lake, Pemadumcook Chain of Lakes, Flagstaff Lake, Mooselookmeguntic Lake, Richardson Lakes (Upper Richardson Lake and Lower Richardson Lake)

Bays and Gulfs: Gulf of Maine, Bay of Fundy, Passamaquoddy Bay, Penobscot Bay, Casco Bay, Merrymeeting Bay

Islands: Mount Desert Island

Archipelagoes: Isles of Shoals

Swamps, Marshes, and Wetlands: Scarborough Marsh

Other Landforms: Bigelow Mountain Ridge

Mountains: Mount Katahdin, Cadillac Mountain, Hamlin Peak, Sugarloaf Mountain, Saddleback Mountain, Old Speck Mountain, Baldpate Mountain

National Parks: Acadia National Park

National Forests: White Mountain National Forest

National Monuments: Katahdin Woods and Waters National Monument

Other Sites: Portland Head Light, Baxter State Park, Quoddy Head State Park, Wells National Estuarine Research Reserve, Seal Island National Wildlife Refuge, Holbrook Island Sanctuary State Park, Portland Observatory, Munjoy's Hill, Old Port, Exchange Street

Dams: Frankfort Dam, Union River Dam, Harris Station Dam

International/Major Airports: Portland International Airport, Bangor International Airport

MARYLAND

Capital: Annapolis

Major Cities: Baltimore, Columbia, Germantown, Silver Spring, Waldorf, Annapolis

Population: 6,020,000

Area: 12,405 sq mi

Bordering States: Delaware, West Virginia, Virginia, Pennsylvania, Washington D.C. (Federal District)

Nicknames: Old Line State, Free State, Chesapeake State, America in Miniature

Mountain Ranges: Appalachian Mountains, Bear Pond Mountains, Allegheny Mountains, Blue Ridge Mountains

Rivers: Susquehanna River, Potomac River, Youghiogheny River, Patuxent River, North Branch Potomac River

Lakes and Reservoirs: Prettyboy Reservoir, Deep Creek Lake, Loch Raven Reservoir, Jennings Randolph Lake

Bays and Gulfs: Chesapeake Bay, Chincoteague Bay (lagoon), Isle of Wight Bay, Mallows Bay, Sinepuxent Bay

Plateaus: Piedmont Plateau, Appalachian Plateau

Peninsulas, Capes, and Points: Delmarva Peninsula, St. Mary's Peninsula, Calvert Peninsula

Islands: Assateague Island

Caves: Crystal Grottoes

Canyons and Gorges: Cumberland Narrows

Swamps, Marshes, and Wetlands: Great Cypress Swamp, Zekiah Swamp, Battle Creek Cypress Swamp

Valleys: Susquehanna Valley, Cumberland Valley

Other Bodies of Water: Intracoastal Waterway, Tangier Sound, Pocomoke Sound, Chesapeake and Ohio Canal, Chesapeake and Delaware Canal

National Monuments: Fort McHenry National Monument, Harriet Tubman Underground Railroad National Monument

Other Sites: Antietam National Battlefield, Assateague Island National Seashore, B&O Railroad Museum, National Aquarium, Washington Monument (Baltimore), Emerson Bromo-Seltzer Tower

Dams: Conowingo Dam

International/Major Airports: Baltimore-Washington International Thurgood Marshall Airport

MASSACHUSETTS

Capital: Boston

Major Cities: Boston, Worcester, Springfield, Lowell, Cambridge

Population: 6,815,000

Area: 10,554 sq mi

Bordering States: Connecticut, Rhode Island, New York, Vermont, New Hampshire

Nicknames: Bay State, Spirit of America, Old Colony State, Pilgrim State

Mountain Ranges: Berkshire Hills, Taconic Mountains, Wapack Range, Metacomet Range, Holyoke Range, Pocumtuck Range

Rivers: Connecticut River, Charles River, Merrimack River, Housatonic River

Lakes and Reservoirs: Quabbin Reservoir, Wachusett Reservoir, Lake Cochituate, Lake Buel, Wenham Lake

Bays and Gulfs: Massachusetts Bay, Buzzards Bay, Cape Cod Bay, Narragansett Bay (small parts of it), Mount Hope Bay, Hingham Bay, Plymouth Bay

Peninsulas, Capes, and Points: Cape Cod, Shawmut Peninsula, Pemberton Point, Deer Island Peninsula, Cape Ann

Isthmuses and Spits: Monomoy Island

Islands: Martha's Vineyard, Plum Island, Spectacle Island, Cuttyhunk Island, Rainsford Island

Archipelagoes: Elizabeth Islands

Caves: Horse Caves

Swamps, Marshes, and Wetlands: Brewster Flats, Hockomock Swamp

Valleys: Connecticut River Valley, Pioneer Valley, Housatonic Valley

Other Landforms: Nantasket Beach

Other Bodies of Water: Nantucket Sound, Vineyard Sound, Provincetown Harbor

Mountains: Mount Greylock, Mount Tom, Mount Holyoke

Other Sites: Cape Cod National Seashore, Stellwagen Bank National Marine Sanctuary, Blue Hills Reservation, Boston Harbor Islands National Recreation Area, Fenway Park, Boston Common, Bunker Hill Monument, Freedom Trail, Boston National Historical Park

Dams: Wachusett Dam

International/Major Airports: Logan International Airport

MICHIGAN

Capital: Lansing

Major Cities: Detroit, Grand Rapids, Warren, Sterling Heights, Ann Arbor, Lansing

Population: 9,930,000

Area: 96,713 sq mi

Bordering States: Wisconsin, Indiana, Ohio

Nicknames: Wolverine State, Great Lakes State

Mountain Ranges: Huron Mountains, Porcupine Mountains

Rivers: Grand River, Muskegon River, St. Joseph River, Manistee River, River Raisin, Au Sable River, Huron River, Kalamazoo River

Lakes and Reservoirs: Lake Superior, Lake Michigan, Lake Huron, Lake Erie, Lake St. Clair

Bays and Gulfs: Green Bay, Grand Traverse Bay, Keweenaw Bay, Saginaw Bay, Little Traverse Bay, Whitefish Bay

Straits: Straits of Mackinac

Peninsulas, Capes, and Points: Upper Peninsula, Lower Peninsula, Keweenaw Peninsula, Leelanau Peninsula, Abbaye Peninsula, Garden Peninsula, Waugoshance Point, The Thumb

Islands: Isle Royale, Mackinac Island, Round Island, Beaver Island, North Manitou Island, South Manitou Island, Bois Blanc Island, Neebish Island

Archipelagoes: Beaver Islands, Fox Islands

Swamps, Marshes, and Wetlands: Lakeville Swamp, Tobico Marsh

Other Landforms: Sleeping Bear Dunes, Grand Sable Dunes, Arch Rock, Nordhouse Dunes, Niagara Escarpment

Other Bodies of Water: Tahquamenon Falls, Munising Fals

Mountains: Mount Arvon

National Parks: Isle Royale National Park

National Forests: Hiawatha National Forest, Huron-Manistee National Forest, Ottawa National Forest

Other Sites: Sleeping Bear Dunes National Seashore, Pictured Rocks National Lakeshore

Dams: Hardy Dam

International/Major Airports: Detroit Metropolitan Wayne County Airport, Gerald R. Fort International Airport, Bishop International Airport

MINNESOTA

Capital: St. Paul

Major Cities: Minneapolis, St. Paul, Rochester, Bloomington, Duluth

Population: 5,520,000

Area: 86,935 sq mi

Bordering States: North Dakota, South Dakota, Iowa, Wisconsin

Nicknames: North Star State, Gopher State, Land of 10,000 Lakes

Mountain Ranges: Sawtooth Mountains, Misquah Hills

Rivers: Mississippi River, Red River of the North, Minnesota River, Wapsipinicon River, Des Moines River, Cedar River, Little Sioux River, Roseau River

Lakes and Reservoirs: Lake Superior, Lake of the Woods, Lake Itasca, Rainy Lake, Mille Lacs Lake, Leach Lake, Lake Vermilion, Lake Pepin, Lake Minnetonka, Lake Nokomis, Lake of the Isles, Cedar Lake, Lake Calhoun, Little Rock Lake, Bay Lake, Big Stone Lake, Bear Island Lake, Otter Tail Lake, Saganaga Lake, Ten Mile Lake, Leech Lake, Lake Bemidji, Lake Winnebigoshish

Isthmuses and Spits: Minnesota Point

Valleys: Red River Valley

Other Landforms: Mesabi Range, Palisade Head, Sugar Loaf, Leaf Hills Moraines

Mountains: Eagle Mountain

National Parks: Voyageurs National Park

National Forests: Chippewa National Forest, Superior National Forest

National Monuments: Grand Portage National Monument, Pipestone National Monument

Other Sites: Minnehaha Park, Mall of America, Gooseberry Falls State Park, U.S. Bank Stadium, Grand Rounds National Scenic Highway, Minneapolis Chain of Lakes Regional Park,

Dams: Sartell Dam, Lake Zumbro Hydroelectric Generating Plant, Deer Lake Dam, Leech Lake Dam, Winnibigoshish Lake Dam

International/Major Airports: Minneapolis-St. Paul International Airport, Duluth International Airport

MISSISSIPPI

Capital: Jackson

Major Cities: Jackson, Gulfport, Southaven, Hattiesburg, Biloxi

Population: 2,990,000

Area: 48,431 sq mi

Bordering States: Alabama, Louisiana, Tennessee, Arkansas

Nicknames: Magnolia State, Birthplace of America's Music

Rivers: Mississippi River, Tennessee River, Pearl River, Tombigbee River, Yazoo River, Chickasawhay River

Lakes and Reservoirs: Pickwick Lake, Grenada Lake, Ross Barnett Reservoir

Bays and Gulfs: Gulf of Mexico

Plains and Basins: Gulf Coastal Plain

Islands: Dauphin Island, Horn Island, Cat Island

Other Bodies of Water: Mississippi Sound

Mountains: Woodall Mountain

National Forests: Bienville National Forest, Delta National Forest, De Soto National Forest, Holly Springs National Forest, Homochitto National Forest, Tombigbee National Forest

Other Sites: Gulf Islands National Seashore, Dauphin Island Bridge

International/Major Airports: Jackson-Evers International Airport, Gulfport-Biloxi International Airport

MISSOURI

Capital: Jefferson City

Major Cities: Kansas City, St. Louis, Springfield, Independence, Columbia, St. Joseph, Joplin, Jefferson City

Population: 6,100,000

Area: 69,706 sq mi

Bordering States: Arkansas, Kansas, Tennessee, Kentucky, Iowa, Illinois, Nebraska, Oklahoma

Nicknames: Show Me State, Bullion State

Mountain Ranges: St. Francois Mountains, Ozark Mountains, Finleys Mountains

Rivers: Missouri River, Mississippi River, White River, St. Francis River, Black River, Des Moines River, Current River, Nishnabotna River, Chariton River, Little Platte River

Lakes and Reservoirs: Truman Reservoir, Lake of the Ozarks, Bull Shoals Lake, Table Rock Lake, Mark Twain Lake, Pomme de Terre Lake, Lake Taneycomo, Stockton Lake,

Plateaus: Ozark Plateau

Plains and Basins: Brazeau Bottom, Gulf Coastal Plain

Caves: Meramec Caverns, Marvel Cave, Mark Twain Cave

Swamps, Marshes, and Wetlands: American Bottom, Bois Brule Bottom, Le Grand Champ Bottom

Valleys: Missouri River Valley, Arcadia Valley

Other Landforms: Tower Rock, Weaubleau Structure, Loess Hills, Decaturville Crater, Pilot Knob

Other Bodies of Water: Big Spring, Greer Spring

Mountains: Taum Sauk Mountain, Hughes Mountain

National Forests: Mark Twain National Forest

National Monuments: George Washington Carver National Monument

Other Sites: Gateway Arch, National World War I Museum and Memorial, Silver Dollar City, Jefferson National Expansion Memorial, Wilson's Creek National Battlefield, Golden Prairie, Elephant Rocks State Park, Grand Gulf State Park, Lake of the Ozarks State Park, Loess Bluffs National Wildlife Refuge

Dams: Table Rock Dam

International/Major Airports: St. Louis Lambert International Airport, Kansas City International Airport, Springfield-Branson National Airport

MONTANA

Capital: Helena

Major Cities: Billings, Missoula, Great Falls, Bozeman, Butte, Helena

Population: 1,050,000

Area: 147,039 sq mi

Bordering States: Idaho, Wyoming, North Dakota, South Dakota

Nicknames: Treasure State, Big Sky Country

Mountain Ranges: Rocky Mountains, Beartooth Mountains, Absaroka Range, Crazy Mountains, Madison Range, Gallatin Range, Mission Mountains, Sapphire Mountains, Anaconda Range, Coeur d'Alene Mountains, Cabinet Mountains, Garnet Range, Mission Mountains, Flint Creek Range, Tobacco Root Mountains, Pryor Mountains, Snowy Mountains

Rivers: Missouri River, Milk River, Yellowstone River, Kootenai River, Bighorn River, Powder River

Lakes and Reservoirs: Flathead Lake, Fort Peck Lake

Plateaus: Beartooth Plateau

Plains and Basins: Great Plains

Caves: Lewis and Clark Caverns

Canyons and Gorges: Bad Rock Canyon

Valleys: Bitterroot Valley, Gallatin Valley, Paradise Valley

Other Landforms: LaBarge Rock, Lewis Overthrust, Iceberg Cirque, Blackfoot Glacier, Sperry Glacier, Grinnell Glacier, Adel Mountains Volcanic Field, Sweet Grass Hills

Other Bodies of Water: Giant Springs

Mountains: Granite Peak, Mount Wood, Hilgard Peak, Crazy Peak, Electric Peak, Emigrant Peak

National Parks: Glacier National Park, Yellowstone National Park

National Forests: Beaverhead-Deerlodge National Forest, Bitterroot National Forest, Custer National Forest, Flathead National Forest, Gallatin National Forest, Helena National Forest, Idaho Panhandle National Forest, Kootenai National Forest, Lewis and Clark National Forest, Lolo National Forest

National Monuments: Little Bighorn Battlefield National Monument, Pompeys Pillar National Monument, Upper Missouri River Breaks National Monument

Other Sites: Going-to-the-Sun Road, Museum of the Rockies, Berkeley Pit, Beartooth Highway

Dams: Yellowtail Dam, Fort Peck Dam, Hungry Horse Dam, Libby Dam, Kerr Dam

International/Major Airports: Bozeman Yellowstone International Airport, Billings Logan International Airport, Missoula International Airport

NEBRASKA

Capital: Lincoln

Major Cities: Omaha, Lincoln, Bellevue, Grand Island, Kearney

Population: 1,910,000

Area: 77,347 sq mi

Bordering States: Kansas, Iowa, Missouri, South Dakota, Colorado, Wyoming

Nicknames: Cornhusker State

Rivers: Missouri River, North Platte River, Niobrara River, Republican River, White River, South Platte River, Elkhorn River, Frenchman Creek, Arikaree River

Lakes and Reservoirs: Harlan County Reservoir, Lake McConaughy, Lewis and Clark Lake, Salt Valley Lakes

Plains and Basins: Great Plains, High Plains, Dissected Till Plains, Nine Hill Prairie, Smoky Hills

Swamps, Marshes, and Wetlands: Rainwater Basin

Valleys: Missouri River Valley, North Platte River Valley, Niobrara River Valley

Other Landforms: Panorama Point, Chimney Rock, Sandhills, Wildcat Hills (escarpment), Pine Ridge, Guide Hill, Courthouse and Jail Rocks

National Forests: Nebraska National Forest, Samuel R. McKelvie National Forest

National Monuments: Agate Fossil Beds National Monument, Homestead National Monument, Scotts Bluff National Monument

Other Sites: Carhenge, Robidoux Pass, The Archway (Great Platte River Road Archway Monument), Pawnee Lake State Recreation Area, Heartland of America Park, Old Market Neighborhood, Oglala National Grassland, Pine Ridge National Recreation Area, Crescent Lake National Wildlife Refuge, DeSoto National Wildlife Refuge, Toadstool Geologic Park, Chadron State Park, Strategic Air Command and Aerospace Museum, Lauritzen Gardens, Chimney Rock National Historic Site, Ash Hollow State Historical Park

Dams: Medicine Creek Dam, Gavins Point Dam, Kingsley Dam

International/Major Airports: Eppley Airfield, Lincoln Airport, Central Nebraska Regional Airport

NEVADA

Capital: Carson City

Major Cities: Las Vegas, Henderson, Reno, North Las Vegas, Sparks, Carson City

Population: 2,950,000

Area: 110,571 sq mi

Bordering States: California, Utah, Arizona, Oregon, Idaho

Nicknames: Silver State

Mountain Ranges: Ruby Mountains, Snake Range, Sierra Nevada, Toiyabe Range, McCullough Range, East Humboldt Range, Spring Mountains, Diamond Mountains

Rivers: Colorado River, Owyhee River, Humboldt River, Amargosa River, Reese River

Lakes and Reservoirs: Lake Tahoe, Lake Mead, Pyramid Lake, Honey Lake, Walker Lake, Lake Mohave

Deserts: Mojave Desert, Amargosa Desert, Black Rock Desert, Smoke Creek Desert, Great Basin Desert

Plateaus: Modoc Plateau

Plains and Basins: Great Basin

Canyons and Gorges: Seitz Canyon

Swamps, Marshes, and Wetlands: Las Vegas Wash, Humboldt Salt Marsh

Valleys: Amargosa Valley, Antelope Valley, Bitter Spring Valley, Pahrump Valley, Eldorado Valley, Piute Valley, Lahontan Valley, Ruby Valley

Other Landforms: Little Finland, Carson Sink, Humboldt Sink, Moenkopi Formation, Devils Hole

Other Bodies of Water: Calville Bay

Mountains: Wheeler Peak, Mount Moriah, Mount Jefferson, Charleston Peak, North Schell Peak, Arc Dome

National Parks: Great Basin National Park

National Forests: Humboldt-Toiyabe National Forest, Inyo National Forest, Lake Tahoe Basin Management Unit

National Monuments: Basin and Range National Monument, Gold Butte National Monument, Tule Springs Fossil Beds National Monument

Other Sites: Red Rock Canyon National Conservation Area, Valley of Fire State Park, Lake Mead National Recreation Area, World Market Center, National Automobile Museum

Dams: Hoover Dam

International/Major Airports: McCarran International Airport, Reno/Tahoe International Airport

NEW HAMPSHIRE

Capital: Concord

Major Cities: Manchester, Nashua, Concord, Derry, Rochester

Population: 1,335,000

Area: 9,349 sq mi

Bordering States: Vermont, Maine, Massachusetts

Nicknames: Granite State, White Mountain State

Mountain Ranges: White Mountains, Presidential Range, Appalachian Mountains, Franconia Range, Sandwich Range, Dartmouth Range, Carter-Moriah Range, Kinsman Range (Cannon-Kinsman Range), Mahoosuc Range

Rivers: Connecticut River, Merrimack River, Saco River, Androscoggin River, Contoocook River, Pemigewasset River

Lakes and Reservoirs: Lake Winnipesaukee, Umbagog Lake, Squam Lake, Newfound Lake, Winnisquam Lake, Lake Sunapee, Moore Reservoir, Ossipee Lake, Lake Wentworth, Massabessic Lake, Lake Francis, Merrymeeting Lake, Great East Lake, Conway Lake, Mascoma Lake

Bays and Gulfs: Paugus Bay, Opechee Bay

Peninsulas, Capes, and Points: Finisterre Point

Islands: Star Island, Rattlesnake Island, Three Mile Island

Archipelagoes: Isles of Shoals

Other Landforms: Crawford Notch, Flume Gorge, The Cannon Balls Mountain Ridge

Other Bodies of Water: Arethusa Falls

Other Landforms: Carter Dome, Pinkham Notch

Mountains: Mount Washington, Mount Adams, Mount Jefferson, Mount Lafayette, South Twin Mountain, Wildcat Mountain, Mount Moosilauke, Mount Carrigain, North Carter Mountain, Middle Carter Mountain, South Carter Mountain, Mount Hight, Franconia Ridge, Mount Garfield,

National Forests: White Mountain National Forest

Dams: Moore Dam

International/Major Airports: Manchester-Boston Regional Airport, Portsmouth International Airport

NEW JERSEY

Capital: Trenton

Major Cities: Newark, Jersey City, Paterson, Elizabeth, Edison

Population: 8,950,000

Area: 8,722 sq mi

Bordering States: Pennsylvania, New York, Delaware

Nicknames: Garden State

Mountain Ranges: Kittatinny Mountains, Ramapo Mountains, Watchung Mountains, Preakness Range, Bearfort Mountain Range

Rivers: Delaware River, Wallkill River, Hudson River, Passaic River, Great Egg Harbor River

Lakes and Reservoirs: Lake Hopatcong, Merrill Creek Reservoir, Monksville Reservoir, Round Valley Reservoir, Manasquan Reservoir

Bays and Gulfs: Delaware Bay, Barnegat Bay, Raritan Bay, Great Bay, Manahawkin Bay

Straits: Kill Van Kull Strait

Plateaus: Hunterdon Plateau, Piedmont Plateau

Plains and Basins: Pine Barrens, Newark Basin, Atlantic Coastal Plain

Peninsulas, Capes, and Points: Cape May, Barnegat Peninsula, Bergen Point, Finns Point

Isthmuses and Spits: Sandy Hook

Islands: Long Beach Island, Absecon Island, Seven Mile Island

Canyons and Gorges: Boonton Gorge

Swamps, Marshes, and Wetlands: Hatfield Swamp, Hackensack Meadowlands, Troy Meadows, Great Swamp

Valleys: Amwell Valley, Kittatinny Valley, Great Appalachian Valley, Wallkill Valley, Wallpack Valley

Other Landforms: The Palisades

Other Bodies of Water: Staten Island Sound, Shark River Inlet, Delaware and Raritan Canal, Intracoastal Waterway

Mountains: High Point

National Monuments: Statue of Liberty National Monument

Other Sites: Liberty State Park, George Washington Bridge, Island Beach State Park, Bayonne Bridge, Stokes State Forest, Lucy the Elephant, Union Watersphere, Lake Carnegie, Thomas Edison National Historical Park, Asbury Park, Princeton Battlefield State Park, Great Swamp National Wildlife Refuge, New Jersey Pinelands National Reserve, Hutcheson Memorial Forest, Bear Swamp, The Glades

International/Major Airports: Newark Liberty International Airport, Atlantic City International Airport

NEW MEXICO

Capital: Santa Fe

Major Cities: Albuquerque, Las Cruces, Rio Rancho, Santa Fe, Roswell

Population: 2,085,000

Area: 121,590 sq mi

Bordering States: Arizona, Utah, Colorado, Texas, Oklahoma

Nicknames: Land of Enchantment

Mountain Ranges: Rocky Mountains, Sangre de Cristo Mountains, Sandia Mountains, Sacramento Mountains, Jemez Mountains, Sierra Blanca, Capitan Mountains, Brazos Mountains, Guadalupe Mountains, Caballo Mountains, Mogollon Mountains, Organ Mountains

Rivers: Rio Grande, Pecos River, Canadian River, Cimarron River, Gila River, San Juan River, Puerco River

Lakes and Reservoirs: Elephant Butte Reservoir, Navajo Lake, Caballo Lake

Deserts: Chihuahuan Desert

Plateaus: Colorado Plateau, Pajarito Plateau, Caja del Rio Plateau, Defiance Plateau

Plains and Basins: Tularosa Basin, Jordana del Muerto Desert Basin, Guzman Basin

Caves: Lechuguilla Cave

Canyons and Gorges: Skeleton Canyon

Valleys: San Luis Valley, Hachita Valley, Mesilla Valley

Other Landforms: Taos Plateau Volcanic Field, Shiprock, Potrillo Volcanic Field, Tooth of Time, Enchanted Mesa, Fajada Butte

Mountains: Wheeler Peak, Truchas Peak, Venado Peak, Santa Fe Baldy, Baldy Mountain, Mount Phillips, Redondo Peak

National Parks: Carlsbad Caverns National Park

National Forests: Apache-Sitgreaves National Forest, Carson National Forest, Cibola National Forest, Coronado National Forest, Gila National Forest, Lincoln National Forest, Santa Fe National Forest

National Monuments: Aztec National Monument, Bandelier National Monument, Capulin Volcano National Monument, El Malpais National Monument, El Morro National Monument, Fort Union National Monument, Gila Cliff Dwellings National Monument, Kasha-Katuwe Tent Rocks National Monument, Organ Mountains-Desert Rocks National Monument, Petroglyph National Monument, Prehistoric Trackways National Monument, Rio Grande del Norte National Monument, Salinas Pueblo Missions National Monument, White Sands National Monument

Other Sites: Chaco Culture National Historical Park, Trails of the Ancients Scenic Byway, Museum of International Folk Art

Dams: Elephant Butte Dam, Navajo Dam

International/Major Airports: Albuquerque International Airport

NEW YORK

Capital: Albany

Major Cities: New York City, Buffalo, Rochester, Yonkers, Syracuse, Albany

Population: 19,775,000

Area: 54,554 sq mi

Bordering States: Connecticut, Pennsylvania, Massachusetts, Vermont, New Jersey

Nicknames: The Empire State

Mountain Ranges: Adirondack Mountains, Catskill Mountains, Hudson Highlands, MacIntyre Mountains, Shawangunk Ridge, Marlboro Mountains

Rivers: Hudson River, St. Lawrence River, Allegheny River, Susquehanna River, Mohawk River, Genesee River, Delaware River

Lakes and Reservoirs: Lake Erie, Lake Ontario, Lake George, Oneida Lake, Finger Lakes (Lake Seneca, Lake Cayuga, Canandaigua Lake, Keuka Lake, Owasco Lake, Skaneateles Lake), Kensico Reservoir, Great Sacandaga Lake, Lake Tear of the Clouds, Roundout Reservoir

Bays and Gulfs: Upper New York Bay, Lower New York Bay, Jamaica Bay

Straits: Buttermilk Channel

Plateaus: Allegheny Plateau

Peninsulas, Capes, and Points: Rockaway Peninsula

Islands: Long Island, Manhattan Island, Staten Island, Ellis Island, Governors Island, Liberty Island, Randalls and Wards Islands

Caves: Howe Caverns

Canyons and Gorges: Niagara Gorge, Ausable Chasm

Swamps, Marshes, and Wetlands: Constitution Marsh

Valleys: Hudson Valley, Wallkill Valley, Great Appalachian Valley

Other Landforms: Niagara Escarpment, Palisades, Onondaga Formation

Other Bodies of Water: Long Island Sound, Niagara Falls, Taughannock Falls

Mountains: Mount Marcy, Algonquin Peak, Mount Haystack

National Forests: Finger Lakes National Forest

National Monuments: African Burial Grounds National Monument, Castle Clinton National Monument, Fort Stanwix National Monument, Governors Island National Monument, Statue of Liberty National Monument, Stonewall National Monument

Other Sites: Cornell Botanical Gardens, Empire State Building, Niagara Falls, Rockefeller Center, Times Square, International United Nations Headquarters, Adirondack Park, Finger Lakes, Broadway Theater, Central Park, Metropolitan Museum of Art, One World Trade Center, Brooklyn Bridge, Unisphere, Chrysler Building, Theodore Roosevelt Birthplace National Historic Site, Federal Hall National Memorial, Belvedere Castle, Washington Square Arch, Butler Library, Carnegie Hall, Big Duck, Chittenango Falls State Park, Saratoga National Historical Park, Federal Hall National Memorial, General Grant National Memorial, Hamilton Grange National Memorial

Dams: Mount Morris Dam, New Croton Dam

International/Major Airports: John F. Kennedy International Airport, LaGuardia Airport, Buffalo Niagara International Airport, Albany International Airport, Greater Rochester International Airport

NORTH CAROLINA

Capital: Raleigh

Major Cities: Charlotte, Raleigh, Greensboro, Durham, Winston-Salem

Population: 10,150,000

Area: 53,819 sq mi

Bordering States: South Carolina, Virginia, Tennessee, Georgia

Nicknames: Tar Heel State, First in Flight State

Mountain Ranges: Great Smoky Mountains, Black Mountains, Unicoi Mountains, Appalachian Mountains, Saluda Mountains, Great Balsam Mountains, Cane Creek Mountains

Rivers: Roanoke River, New River, Neuse River, Catawba River

Lakes and Reservoirs: Lake Norman, Mountain Island Lake, Badin Lake

Plateaus: Piedmont Plateau

Peninsulas, Capes, and Points: Cape Hatteras, Albemarle-Pamlico Peninsula, Cape Lookout, Cape Fear, Bodie Island Barrier Peninsula

Islands: Hatteras Island, Ocracoke Island, Roanoke Island, Masonboro Island

Archipelagoes: Outer Banks, Core Banks

Swamps, Marshes, and Wetlands: Great Dismal Swamp

Valleys: Tennessee Valley

Other Landforms: Crystal Coast

Other Bodies of Water: Albemarle Sound, Pamlico Sound, Currituck Sound, Oregon Inlet, Masonboro Inlet

Mountains: Mount Mitchell, Mount Craig, Clingmans Dome, Pilot Mountain, Waterrock Knob, Mount Pisgah, Craggy Dome

National Parks: Great Smoky Mountains National Park

National Forests: Cherokee National Forest, Croatan National Forest, Nantahala National Forest, Pisgah National Forest, Uwharrie National Forest

Other Sites: Cape Hatteras National Seashore, Cape Lookout National Seashore, Wright Brothers National Memorial, Chimney Rock State Park, Fort Raleigh National Historic Site, Moores Creek National Battlefield

Dams: Fontana Dam

International/Major Airports: Charles/Douglas International Airport, Raleigh-Durham International Airport

NORTH DAKOTA

Capital: Bismarck

Major Cities: Fargo, Bismarck, Grand Forks, Minot, West Fargo

Population: 757,000

Area: 70,698 sq mi

Bordering States: South Dakota, Montana, Minnesota

Nicknames: Peace Garden State

Rivers: Missouri River, James River, Yellowstone River, Red River of the North, Sheyenne River, Little Missouri River

Lakes and Reservoirs: Lake Sakakawea, Devils Lake

Plateaus: Missouri Plateau, Turtle Mountain Plateau

Plains and Basins: Great Plains

Valleys: Red River Valley

Other Landforms: White Butte, Pembina Escarpment

National Parks: Theodore Roosevelt National Park

Other Sites: International Peace Garden, Ralph Engelstad Arena, Scandinavian Heritage Park, World's Largest Buffalo Monument

International/Major Airports: Hector International Airport

OHIO

Capital: Columbus

Major Cities: Columbus, Cleveland, Cincinnati, Toledo, Akron

Population: 11,615,000

Area: 44,825 sq mi

Bordering States: Indiana, Kentucky, Pennsylvania, West Virginia, Michigan

Nicknames: Buckeye State

Mountain Ranges: Appalachian Mountains

Rivers: Ohio River, Wabash River, Scioto River, Olentangy River, Great Miami River, Maumee River, Cuyahoga River, Sandusky River, Muskingum River, Mahoning River

Lakes and Reservoirs: Lake Erie, Pymatuning Reservoir, Grand Lake St. Marys, Portage Lakes, Tappan Lake

Bays and Gulfs: Maumee Bay, Sandusky Bay

Plateaus: Allegheny Plateau

Peninsulas, Capes, and Points: Whiskey Island

Islands: Kelleys Island, South Bass Island, Gulf Island Shoal

Archipelagoes: Lake Erie Islands, Bass Islands

Caves: Ohio Caverns, Seneca Caverns

Canyons and Gorges: Blackhand Gorge

Swamps, Marshes, and Wetlands: Great Black Swamp

Valleys: Cuyahoga Valley, Miami Valley

Other Landforms: Campbell Hill, Hocking Hills, Mississinawa Moraine

National Parks: Cuyahoga Valley National Park

National Forests: Wayne National Forest

National Monuments: Charles Young Buffalo Soldiers National Monument

Other Sites: Grand Lake St. Marys State Park, Perry's Victory and International Peace Memorial, Hopewell Culture National Historical Park, Cedar Point, Hocking Hills State Park, Blackhand Gorge State Nature Preserve, Rock and Roll Hall of Fame, Key Tower, Fountain of Eternal Life (War Memorial Fountain), James A. Garfield Memoial

Dams: Mohawk Dam

International/Major Airports: Cleveland-Hopkins International Airport, Port Columbus International Airport

OKLAHOMA

Capital: Oklahoma City

Major Cities: Oklahoma City, Tulsa, Norman, Broken Arrow, Lawton, Edmond, Moore

Population: 3,925,000

Area: 69,898 sq mi

Bordering States: Texas, Kansas, Arkansas, Colorado, Missouri, New Mexico

Nicknames: Sooner State, Native America

Mountain Ranges: Ouachita Mountains, Ozark Mountains, Wichita Mountains, Arbuckle Mountains, Boston Mountains, Kiamichi Mountains, Quartz Mountains

Rivers: Arkansas River, Red River of the South, Canadian River, Cimarron River, Neosho River, North Canadian River, Verdigris River, Washita River, Kiamichi River

Lakes and Reservoirs: Eufaula Lake, Lake Texoma, Grand Lake of the Cherokee, Robert S. Kerr Reservoir, Oologah Lake, Keystone Lake, Sardis Lake, Tenkiller Ferry Lake, Broken Bow Lake, Skiatook Lake, Great Salt Plains Lake, Lake of the Arbuckles

Plateaus: Ozark Plateau

Plains and Basins: Gulf Coastal Plain, Great Plains, Johns Valley

Caves: Alabaster Caverns

Other Landforms: Black Mesa, Cavanal Hill, Glass Mountains, Antelope Hills, Mesa de Maya, Flint Hills, Ames Crater

Other Bodies of Water: Turner Falls

Mountains: Mount Scott, Mount Pinchot, Rich Mountain, Lookout Mountain, Quartz Mountain

National Forests: Ouachita National Forest

Other Sites: Tallgrass Prairie Preserve, Black Kettle National Grassland, Fort Smith National Historic Site, Oklahoma City National Memorial, Chickasaw National Recreation Area, Golden Driller, Myriad Botanical Gardens, National Cowboy and Western Heritage Museum, Ed Galloway's Totem Pole Park, Blue Whale of Catoosa, Great Salt Plains State Park, Talimena Scenic Drive, Wichita Mountains Wildlife Refuge, Washita Battlefield National Historic Site, BOK Tower, Philbrook Museum of Art

Dams: Pensacola Dam

International/Major Airports: Will Rogers World Airport, Tulsa International Airport

OREGON

Capital: Salem

Major Cities: Portland, Salem, Eugene, Gresham, Hillsboro

Population: 4,095,000

Area: 98,378 sq mi

Bordering States: California, Washington, Idaho, Nevada

Nicknames: Beaver State

Mountain Ranges: Cascade Range, Coast Range, Klamath Mountains, Siskiyou Mountains, Blue Mountains, Pueblo Mountains, Wallowa Mountains, Calapooya Mountains

Rivers: Columbia River, Snake River, Owyhee River, John Day River, Klamath River, Deschutes River, Rogue River, Malheur River, Willamette River, Grande Ronde River

Lakes and Reservoirs: Crater Lake

Bays and Gulfs: Netarts Bay, Tillamook Bay, Yaquina Bay

Deserts: Alvord Desert, High Desert, Owyhee Desert

Plateaus: Columbia Plateau, Modoc Plateau

Plains and Basins: Klamath Basin

Peninsulas, Capes, and Points: Cape Lookout, Cascade Head, Cape Meares

Islands: Zwagg Island

Archipelagoes: Oregon Islands

Caves: Lava River Cave, Redmond Caves, Paisley Caves, Derrick Cave, Sea Lion Caves

Canyons and Gorges: Columbia River Gorge, Hells Canyon

Valleys: Willamette Valley, Catlow Valley, Goose Lake Valley

Other Landforms: Three Arch Rocks, Cascade Volcanoes, Pilot Butte, Three Sisters, Broken Top, Newberry Volcano, Horse Lava Tube System

Other Bodies of Water: Willamette Falls, Columbia River Estuary

Mountains: Mount Hood, Mount Jefferson, Neahkahnie Mountain, Mount Bachelor

National Parks: Crater Lake National Park

National Forests: Deschutes National Forest, Fremont-Winema National Forest, Klamath National Forest, Malheur National Forest, Mount Hood National Forest, Ochoco National Forest, Siuslaw National Forest, Wallowa-Whitman National Forest, Willamette National Forest

National Monuments: John Day Fossil Beds National Monument, Newberry National Volcanic Monument, Oregon Caves National Monument

Other Sites: Nez Perce National Historical Park, Fort Vancouver National Historic Site

Dams: Bonneville Dam

International/Major Airports: Portland International Airport

PENNSYLVANIA

Capital: Harrisburg

Major Cities: Philadelphia, Pittsburgh, Allentown, Erie, Reading, Scranton, Lancaster, Bethlehem

Population: 12,815,000

Area: 46,054 sq mi

Bordering States: New York, Ohio, West Virginia, Maryland, Delaware, New Jersey

Nicknames: Keystone State, Quaker State

Mountain Ranges: Appalachian Mountains, Pocono Mountains, Allegheny Mountains, Blue Ridge Mountains, Bear Pond Mountains, Endless Mountains, Conewago Mountains, Moosic Mountains

Rivers: Susquehanna River, Allegheny River, Ohio River, Monongahela River, Delaware River, Schuylkill River, Genesee River, Youghiogheny River

Lakes and Reservoirs: Lake Erie, Lake Wallenpaupack, Raystown Lake, Lake Nockamixon, Pymatuning Reservoir

Bays and Gulfs: Presque Isle Bay

Plateaus: Allegheny Plateau

Caves: Indian Echo Caverns, Lincoln Caverns, Crystal Cave

Canyons and Gorges: Pine Creek Gorge

Swamps, Marshes, and Wetlands: Green Pond Marsh, Espy Bog

Valleys: Susquehanna Valley, Great Appalachian Valley, Cumberland Valley, Kishacoquillas Valley, Bald Eagle Valley, Nittany Valley

Other Landforms: Indian God Rock, Tuscarora Formation, South Mountain, Mount Nittany, Blue Knob, Camelback Mountain, Pine Knob

Mountains: Mount Davis, Big Mountain, Schaefer Head, Bald Mountain, Elk Hill (North Knob or Elk Mountain)

National Forests: Allegheny National Forest

Other Sites: Tioga State Forest, National Constitution Center, Independence Hall, Benjamin Franklin National Memorial, Valley Forge National Historical Park, Buchanan State Forest, Gettysburg National Military Park, Gettysburg National Cemetery, Antietam National Battlefield, Independence Mall, Liberty Bell, Elk Mountain Ski Area, Tomb of the Unknown Revolutionary War Soldier, Washington Square

Dams: Kinzua Dam

International/Major Airports: Philadelphia International Airport, Pittsburgh International Airport, Harrisburg International Airport

RHODE ISLAND

Capital: Providence

Major Cities: Providence, Warwick, Cranston, Pawtucket, East Providence

Population: 1,060,000

Area: 1,544 sq mi

Bordering States: Connecticut, Massachusetts

Nicknames: Ocean State

Rivers: Blackstone River, Pawcatuck River, Wood River

Lakes and Reservoirs: Scituate Reservoir, Ninigret Pond

Bays and Gulfs: Narragansett Bay, Little Narragansett Bay

Peninsulas, Capes, and Points: Quonset Point

Islands: Aquidneck Island, Conanicut Island, Block Island, Prudence Island, Patience Island

Valleys: Blackstone Valley

Other Landforms: Mohegan Bluffs

Other Bodies of Water: Block Island Sound

Mountains: Jerimoth Hill

Other Sites: Roger Williams National Memorial, Blackstone River Valley National Historical Park

International/Major Airports: Theodore Francis Green State Airport

SOUTH CAROLINA

Capital: Columbia

Major Cities: Columbia, Charleston, North Charleston, Mount Pleasant, Rock Hill

Population: 4,975,000

Area: 32,020 sq mi

Bordering States: North Carolina, Georgia

Nicknames: Palmetto State

Mountain Ranges: Appalachian Mountains, Blue Ridge Mountains, Saluda Mountains

Rivers: Savannah River, Catawba River, Edisto River, Saluda River, Black River

Lakes and Reservoirs: Lake Marion, Lake Strom Thurmond (Clarks Hill Lake), Lake Moultrie, Lake Hartwell, Lake Murray, Richard B. Russell Lake, Lake Keowee, Lake Wylie

Bays and Gulfs: Winyah Bay

Peninsulas, Capes, and Points: Waccamaw Neck

Plateaus: Piedmont Plateau

Plains and Basins: Atlantic Coastal Plain, ACE (Ashepoo, Combahee, and Edisto) Basin

Islands: Hilton Head Island, Johns Island, Saint Helena Island, Sullivan's Island, Kiawah Island, Fripp Island, Hunting Island, Callawassie Island

Archipelagoes: Sea Islands

Swamps, Marshes, and Wetlands: Audobon Swamp Garden, Four Holes Swamp

Valleys: Great Appalachian Valley

Other Bodies of Water: Port Royal Sound, Intracoastal Waterway, Charleston Harbor, Saint Helena Sound, Floridan Aquifer

Mountains: Sassafras Mountain, Pinnacle Mountain, Glassy Mountain

National Parks: Congaree National Park

National Forests: Francis Marion National Forest, Sumter National Forest

National Monuments: Fort Sumter National Monument, Reconstruction Era National Monument

Other Sites: Francis Beidler Forest, Cypress Gardens, Charles Pinckney National Historic Site, Table Rock State Park, Caesars Head State Park, Fort Moultrie, Cowpens National Battlefield, Kings Mountain National Military Park, The Pink House, Grand Strand, Brookgreen Gardens, Huntington Beach State Park

Dams: Richard B. Russell Dam, Hartwell Dam, J. Strom Thurmond Dam

International/Major Airports: Charleston International Airport, Greenville-Spartanburg International Airport

SOUTH DAKOTA

Capital: Pierre

Major Cities: Sioux Falls, Rapid City, Aberdeen, Brookings, Watertown

Population: 870,000

Area: 77,115 sq mi

Bordering States: North Dakota, Montana, Minnesota, Iowa, Nebraska, Wyoming

Nicknames: Mount Rushmore State, Blizzard State, Coyote State, Artesian State

Mountain Ranges: Black Hills, Elk Mountains, Bear Lodge Mountains

Rivers: Missouri River, James River, White River, Little Missouri River, Big Sioux River, Cheyenne River, Belle Fourche River

Lakes and Reservoirs: Lake Oahe, Lake Francis Case, Lake Sharpe, Lewis and Clark Lake, Big Stone Lake, Lake Kampeska, Lake Thompson, Pactola Lake, Sylvan Lake, Sheridan Lake

Plateaus: Missouri Plateau, Coteau des Prairies

Plains and Basins: Great Plains, Missouri River Basin

Caves: Wind Cave, Jewel Cave, Rushmore Cave, Sitting Bull Crystal Caverns

Canyons and Gorges: Spearfish Canyon

Other Landforms: Bear Butte, Thunder Butte, Needles (Black Hills)

Mountains: Black Elk Peak, Mount Rushmore, Terry Peak, Cicero Peak, Red Shirt Table Mountain

National Parks: Badlands National Park, Wind Cave National Park

National Forests: Black Hills National Forest, Custer National Forest

National Monuments: Jewel Cave National Monument

Other Sites: Mount Rushmore National Memorial, Missouri National Recreational River, Custer State Park, Crazy Horse Memorial, Mammoth Site, Homestake Gold Mine, Minuteman Missile National Historic Site, Dinosaur Park, Corn Palace, Buffalo Gap National Grassland, Lewis and Clark National Historic Trail, Grand River National Grassland, Washington Pavilion of Arts and Science, Reptile Gardens, Chapel in the Hills

Dams: Oahe Dam, Big Bend Dam, Pactola Dam, Deerfield Dam

International/Major Airports: Sioux Falls Regional Airport

TENNESSEE

Capital: Nashville

Major Cities: Memphis, Nashville, Knoxville, Chattanooga, Clarksville

Population: 6,655,000

Area: 42,144 sq mi

Bordering States: Kentucky, Missouri, Alabama, North Carolina, Virginia, Georgia, Arkansas, Mississippi

Nicknames: Volunteer State, Butternut State, Big Bend State

Mountain Ranges: Great Smoky Mountains, Unicoi Mountains, Appalachian Mountains, Unaka Mountains, Bald Mountains

Rivers: Tennessee River, Cumberland River, Mississippi River, Clinch River, Duck River, Hatchie River, French Broad River, Elk River, Holston River, Powell River

Lakes and Reservoirs: Kentucky Lake, Guntersville Lake, Wheeler Lake, Lake Barkley, Pickwick Lake

Plateaus: Cumberland Plateau

Plains and Basins: Gulf Coastal Plain, Nashville Basin

Caves: Cumberland Caverns, Craighead Caverns

Canyons and Gorges: Scott's Gulf

Valleys: Great Appalachian Valley, Tennessee Valley, The Sugarlands

Other Landforms: Walden Ridge

Mountains: Mount Guyot, Mount Le Conte, Mount Chapman, Mount Kephart, Thunderhead Mountain, Big Frog Mountain

National Parks: Great Smoky Mountains National Park

National Forests: Cherokee National Forest, Land Between the Lakes National Forest

Other Sites: Memphis Pyramid, Sunsphere, Cumberland Gap National Historical Park, Andrew Johnson National Historic Site, Chickamauga and Chattanooga National Military Park, Shiloh National Military Park, Cherohala Skyway, Joyce Kilmer Memorial Forest

Dams: Watauga Dam, Calderwood Dam

International/Major Airports: Memphis International Airport, Nashville International Airport

TEXAS

Capital: Austin

Major Cities: Houston, Dallas, San Antonio, Austin, Fort Worth, El Paso, Arlington, Corpus Christi, Plano, Laredo, Lubbock, Garland, Irving, Amarillo, Grand Prairie, Waco, Abilene, Odessa, Brownsville, Pasadena

Population: 27,875,000

Area: 268,596 sq mi

Bordering States: New Mexico, Oklahoma, Arkansas, Louisiana

Nicknames: Lone Star State, Friendship State

Mountain Ranges: Guadalupe Mountains, Davis Mountains, Chisos Mountains, Franklin Mountains

Rivers: Rio Grande, Red River, Pecos River, Brazos River, Colorado River, Canadian River, Trinity River, Sabine River, Neches River, Nueces River

Lakes and Reservoirs: Toledo Bend Reservoir, Sam Rayburn Reservoir, Falcon International Reservoir, Lake Texoma, Lake Livingston, Amistad Reservoir, Green Lake, Caddo Lake

Bays and Gulfs: Gulf of Mexico, Galveston Bay, Corpus Christi Bay, Nueces Bay, Oso Bay, Matagorda Bay, San Antonio Bay, Aransas Bay, Copano Bay, Lavaca Bay, Carancahua Bay

Straits: San Luis Pass

Deserts: Chihuahuan Desert

Plateaus: Edwards Plateau

Plains and Basins: Great Plains, Gulf Coastal Plain

Peninsulas, Capes, and Points: Bolivar Peninsula

Islands: Padre Island, Galveston Island, Mustang Island, San José Island

Caves: Natural Bridge Caverns, Caverns of Sonora

Canyons and Gorges: Blanco Canyon, Palo Duro Canyon, McKittrick Canyon

Valleys: Rio Grande Valley, Mesilla Valley

Other Landforms: Caprock Escarpment, Mescalero Escarpment, Llano Uplift, Balcones Fault

Other Bodies of Water: Edwards Aquifer, San Marcos Springs

Mountains: Guadalupe Peak, Bush Mountain, Shumard Peak, Baldy Peak, Hunter Peak, El Capitan, Emory Peak

National Parks: Big Bend National Park, Guadelupe Mountains National Park

National Forests: Angelina National Forest, Davy Crockett National Forest, Sabine National Forest, Sam Houston National Forest

National Monuments: Alibates Flint Quarries National Monument, Military Working Dog Teams National Monument, Waco Mammoth National Monument

Other Sites: Johnson Space Center, Padre Island National Seashore, Caprock Canyons State Park and Trailway, Big Thicket National Preserve, Fort Davis National Historic Site, Amistad National Recreation Area, The Alamo Mission, Globe Life Park, Texas State Park

Dams: Mansfield Dam

International/Major Airports: Dallas/Fort Worth International Airport, George Bush Intercontinental Airport, William P. Hobby International Airport, Austin-Bergstrom International Airport, El Paso International Airport

UTAH

Capital: Salt Lake City

Major Cities: Salt Lake City, West Valley City, Provo, West Jordan, Orem

Population: 3,055,000

Area: 84,896 sq mi

Bordering States: Nevada, Arizona, Wyoming, Colorado, New Mexico, Idaho

Nicknames: Beehive State, Mormon State

Mountain Ranges: Rocky Mountains, Wasatch Range, Uinta Mountains, Oquirrh Mountains, Bear River Mountains, Stansbury Mountains, La Sal Mountains, Deep Creek Mountains, Tushar Mountains, Henry Mountains

Rivers: Colorado River, Green River, Bear River, San Juan River, Sevier River, Dolores River

Lakes and Reservoirs: Great Salt Lake, Lake Powell, Utah Lake, Sevier Lake

Deserts: Great Basin Desert, Ferguson Desert, Sevier Desert

Plateaus: Colorado Plateau, Markagunt Plateau, Kaiparowits Plateau, Aquarius Plateau, Paunsaugunt Plateau

Plains and Basins: Great Basin, Uintah Basin

Peninsulas, Capes, and Points: Promontory Point

Islands: Antelope Island

Caves: Timpanogos Cave System

Canyons and Gorges: Zion Canyon, Cataract Canyon, Nine Mile Canyon, Buckskin Gulch, Bryce Canyon, Horseshoe Canyon

Valleys: Salt Lake Valley, Utah Valley, Tule Valley, Wah Wah Valley, Snake Valley, Antelope Valley, Pine Valley

Other Landforms: Bonneville Salt Flats, Rainbow Bridge, Landscape Arch, Kolob Arch, Moenkopi Formation, Grand Staircase, Devil's Garden, Cedar Mesa, Upheaval Dome, Cutler Formation

Mountains: Mount Timpanogos

National Parks: Arches National Park, Bryce Canyon National Park, Capitol Reef National Park, Canyonlands National Park, Zion National Park

National Forests: Ashley National Forest, Dixie National Forest, Fishlake National Forest, Manti-La Sal National Forest, Sawtooth National Forest, Uinta-Wasatch-Cache National Forest

National Monuments: Bears Ears National Monument, Cedar Breaks National Monument, Dinosaur National Monument, Grand Staircase-Escalante National Monument, Hovenweep National Monument, Natural Bridges National Monument, Rainbow Bridge National Monument, Timpanogos Cave National Monument

Other Sites: Temple Square, Antelope Island State Park, Golden Spike National Historic Site

Dams: Flaming Gorge Dam

International/Major Airports: Salt Lake City International Airport

VERMONT

Capital: Montpelier

Major Cities: Burlington, Essex, South Burlington, Colchester, Rutland

Population: 625,000

Area: 9,616 sq mi

Bordering States: New York, New Hampshire, Massachusetts

Nicknames: Green Mountain State

Mountain Ranges: Green Mountains, Taconic Mountains, Notre Dame Mountains

Rivers: Connecticut River, Otter Creek, Winooski River

Lakes and Reservoirs: Lake Champlain, Lake Memphremagog, Lake Bomoseen, Lake Seymour

Bays and Gulfs: Missisquoi Bay

Islands: Isle La Motte

Canyons and Gorges: Quechee Gorge

Valleys: Connecticut River Valley, Champlain Valley

Other Landforms: Green Mountain Giant

Mountains: Mount Mansfield, Killington Peak, Mount Ellen, Jay Peak,

National Forests: Green Mountains National Forest

Dams: Moore Dam

International/Major Airports: Burlington International Airport

VIRGINIA

Capital: Richmond

Major Cities: Virginia Beach, Norfolk, Chesapeake, Arlington, Richmond

Population: 8,415,000

Area: 42,744 sq mi

Bordering States: West Virginia, North Carolina, Maryland, Kentucky, Tennessee, Washington D.C. (Federal District)

Nicknames: The Old Dominion

Mountain Ranges: Appalachian Mountains, George Washington Mountains, Blue Mountains, Cumberland Mountains, Southwest Mountains

Rivers: Roanoke River, James River, New River, Potomac River, Clinch River, Dan River, Rappahannock River

Lakes and Reservoirs: Smith Mountain Lake, Kerr Lake, Lake Gaston, Lake Anna, South Holston Lake, Lake Drummond

Bays and Gulfs: Chesapeake Bay, Chincoteague Bay, Belmont Bay

Straits: Assateague Channel, Chincoteague Channel

Plateaus: Cumberland Plateau, Appalachian Plateau

Plains and Basins: Atlantic Coastal Plain, Tidewater

Peninsulas, Capes, and Points: Delmarva Peninsula, Virginia Peninsula, Middle Peninsula, Northern Neck, Sewells Point, Mason Neck

Isthmuses and Spits: Willoughby Spit

Islands: Chincoteague Island, Jamestown Island, Wallops Island

Archipelagoes: Virginia Barrier Islands

Canyons and Gorges: Mather Gorge

Swamps, Marshes, and Wetlands: Great Dismal Swamp, Dyke Marsh

Valleys: Roanoke Valley, Tennessee Valley, Shenandoah Valley, Great Appalachian Valley

Other Landforms: Natural Bridge, Trimble Knob, Mole Hill, Massanutten Mountain

Mountains: Mount Rogers, Whitetop Mountain, Old Rag Mountain, Hawksbill Mountain

National Parks: Shenandoah National Park

National Forests: George Washington and Jefferson National Forest

National Monuments: Booker T. Washington National Monument, Fort Monroe National Monument, George Washington Birthplace National Monument

Other Sites: Monticello, Assateague Island National Seashore, Blue Ridge Parkway, Arlington National Cemetery, Appomattox Court House, Harpers Ferry National Historical Park, Manassas National Battlefield Park, Fredericksburg National Cemetery, Petersburg National Battlefield, Mid-Atlantic Regional Spaceport

Dams: John H. Kerr Dam

International/Major Airports: Washington Dulles International Airport, Ronald Reagan International Airport, Norfolk International Airport, Richmond International Airport

WASHINGTON

Capital: Olympia

Major Cities: Seattle, Spokane, Tacoma, Vancouver, Bellevue

Population: 7,290,000

Area: 71,297 sq mi

Bordering States: Oregon, Idaho

Nicknames: Evergreen State

Mountain Ranges: Columbia Mountains, Selkirk Mountains, Cariboo Mountains, Monashee Mountains, Purcell Mountains, Chuckanut Mountains

Rivers: Columbia River, Snake River, Yakima River, Grande Ronde River, Kettle River, Palouse River, Skagit River

Lakes and Reservoirs: Lake Chelan, Lake Washington, Franklin Delano Roosevelt Lake, Potholes Reservoir

Bays and Gulfs: Padilla Bay, Skagit Bay, Willapa Bay, Discovery Bay, Bellingham Bay, Oakland Bay

Straits: Strait of Juan de Fuca, Strait of Georgia, Rosario Strait, Haro Strait, Rich Passage, Deception Pass, Swinomish Channel

Plateaus: Columbia Plateau, Okanagan Highland

Peninsulas, Capes, and Points: Olympic Peninsula, Cape Flattery, Cape Alava, Long Beach Peninsula, Kitsap Peninsula, Key Peninsula, Quimper Peninsula, Lummi Peninsula

Islands: Orcas Island, San Juan Island, Lopez Island, Whidbey Island, Camano Island, Shaw Island, Fidalgo Island, Guemes Island, Lummi Island, Fox Island

Archipelagoes: San Juan Islands

Canyons and Gorges: Columbia River Gorge, Ape Canyon

Swamps, Marshes, and Wetlands: Saltese Flats

Other Landforms: Carbon Glacier, Emmons Glacier, Cascade Volcanic Arc, The Brothers

Other Bodies of Water: Puget Sound, Admiralty Inlet, Salish Sea, Saratoga Passage, Hammersley Inlet, Olympic Hot Springs, Hood Canal

Mountains: Mount Rainier, Crystal Mountain, Mount St. Helens, Mount Baker, Mount Adams, Glacier Peak (Dakobed), Mount Shuksan, Mount Olympus, Mount Constance, Mount Stuart, Mount Deception

National Parks: Mount Rainier National Park, North Cascades National Park, Olympic National Park

National Forests: Colville National Forest, Gifford Pinchot National Forest, Idaho Panhandle National Forest, Mount Baker-Snoqualmie National Forest, Okanogan-Wenatchee National Forest, Olympic National Forest, Umatilla National Forest

National Monuments: Hanford Reach National Monument, Mount St. Helens National Volcanic Monument, San Juan Islands National Monument

Other Sites: Seattle Space Needle, CenturyLink Field, Lake Union Park, Pike Place Market, Museum of Glass, San Juan Island National Historical Park, Lake Roosevelt National Recreation Area, Deception Pass Bridge, Cape Disappointment State Park

Dams: Grand Coulee Dam, Mossyrock Dam, Ross Dam

International/Major Airports: Seattle-Tacoma International Airport, Spokane International Airport

WEST VIRGINIA

Capital: Charleston

Major Cities: Charleston, Huntington, Morgantown, Parkersburg, Wheeling

Population: 1,835,000

Area: 24,230 sq mi

Bordering States: Ohio, Virginia, Pennsylvania, Maryland, Kentucky

Nicknames: Mountain State, Panhandle State

Mountain Ranges: Appalachian Mountains, Allegheny Mountains, Cumberland Mountains

Rivers: Ohio River, New River, Potomac River, Greenbrier River, Elk River, Little Kanawha River, Guyandotte River

Lakes and Reservoirs: Summersville Lake, Bluestone Lake, Burnsville Lake, Sutton Lake

Plateaus: Appalachian Plateau, Allegheny Plateau, Cumberland Plateau

Islands: Blennerhassett Island

Caves: Lost World Caverns, Organ Cave, Hellhole Pit Cave

Canyons and Gorges: Smoke Hole Canyon, Cheat Canyon

Swamps, Marshes, and Wetlands: Cranberry Glades

Valleys: Canaan Valley, Germany Valley, Ohio River Valley

Other Landforms: Seneca Rocks, Champe Rocks, Spruce Knob, North Fork Mountain, Wills Mountain

Other Bodies of Water: Blackwater Falls

Mountains: Spruce Mountain, Cheat Mountain, Back Allegheny Mountain

National Forests: George Washington and Jefferson National Forest, Monongahela National Forest

Other Sites: New River Gorge Bridge, New River Gorge National River, Ohio River Islands National Wildlife Refuge, Dolly Sods Wilderness, Chesapeake and Ohio Canal National Historical Park, Harpers Ferry National Historical Park, Hawks Nest State Park, Veterans Memorial Bridge, Harris Riverfront Park

Dams: Summersville Dam

International/Major Airports: Yeager Airport

WISCONSIN

Capital: Madison

Major Cities: Milwaukee, Madison, Green Bay, Kenosha, Racine

Population: 5,780,000

Area: 65,496 sq mi

Bordering States: Michigan, Minnesota, Iowa, Illinois

Nicknames: Badger State, America's Dairyland

Mountain Ranges: Baraboo Range

Rivers: Wisconsin River, Mississippi River, Rock River, Wolf River, Oconto River, Pecatonica River, Saint Louis River, Black River, Chippewa River, St. Croix River

Lakes and Reservoirs: Lake Superior, Lake Michigan, Lake Winnebago, Lake Mendota, Lake Monona, Lake Pepin, Petenwell Lake, Lake Wingra, Castle Rock Lake, Lake Chippewa, Lake Poygan, Lake Koshkonong, Shawano Lake

Bays and Gulfs: Green Bay, Chequamegon Bay, Sturgeon Bay, Superior Bay, Oronto Bay, Pokegama Bay

Straits: Porte des Morts Strait

Plains and Basins: Central Plain

Peninsulas, Capes, and Points: Door Peninsula, Bayfield Peninsula, Chequamegon Point

Islands: Washington Island

Caves: Cave of the Mounds, Crystal Cave

Swamps, Marshes, and Wetlands: Coffee Swamp, Theresa Marsh, Horicon Marsh

Valleys: Menomonee Valley

Other Landforms: Timms Hill, Rib Mountain, Niagara Escarpment, Driftless Area

National Forests: Chequamegon-Nicolet National Forest

Other Sites: Apostle Islands National Lakeshore, Ice Age National Scenic Trail, Cherney Maribel Caves County Park, North Point Water Tower, Lakeshore State Park, Miller Park, Gesu Church (Milwaukee), Lake Kegonsa State Park, Monona Terrace Community and Convention Center

Dams: Mondeaux Dam Recreation Area

International/Major Airports: General Mitchell International Airport, Green Bay-Austin Straubel International Airport, Appleton International Airport, Dane County Regional Airport, Central Wisconsin Airport

WYOMING

Capital: Cheyenne

Major Cities: Cheyenne, Casper, Laramie, Gillette, Rock Springs

Population: 590,000

Area: 97,813 sq mi

Bordering States: Idaho, Montana, Colorado, South Dakota, Nebraska, Utah

Nicknames: Cowboy State, Equality State

Mountain Ranges: Rocky Mountains, Teton Range, Wind River Range, Absaroka Range, Bighorn Range, Laramie Mountains, Medicine Bow Mountains, Black Hills

Rivers: Snake River, Green River, North Platte River, Yellowstone River, Niobrara River, Little Missouri River, Bear River, Bighorn River, Powder River, Cheyenne River

Lakes and Reservoirs: Yellowstone Lake, Flaming Gorge Reservoir, Jackson Lake, Jenny Lake

Deserts: Red Desert

Plateaus: Yellowstone Plateau, Death Canyon Shelf

Plains and Basins: Great Divide Basin, Great Plains

Caves: Tongue River Cave

Canyons and Gorges: Grand Canyon of the Yellowstone, Cascade Canyon, Webb Canyon, Garnet Canyon

Valleys: Star Valley, Jackson Hole (Valley), Hayden Valley

Other Landforms: Devils Tower, Teapot Rock

Other Bodies of Water: Yellowstone Falls, Old Faithful Geyser

Mountains: Gannett Peak, Grand Teton, Fremont Peak, Downs Mountain, Wind River Peak, Cloud Peak

National Parks: Grand Teton National Park, Yellowstone National Park

National Forests: Ashley National Forest, Bighorn National Forest, Black Hills National Forest, Bridger-Teton National Forest, Caribou-Targhee National Forest, Medicine Bow-Routt National Forest, Shoshone National Forest, Uinta-Wasatch-Cache National Forest

National Monuments: Devils Tower National Monument, Fossil Butte National Monument

Other Sites: John D. Rockefeller Jr. Memorial Parkway, Bighorn Canyon National Recreation Area

Dams: Pathfinder Dam, Jackson Lake Dam, Buffalo Bill Dam

International/Major Airports: Jackson Hole Airport

United States Territory Fact Files

PUERTO RICO

Capital: San Juan

Major Cities: San Juan, Ponce, Bayamon, Carolina, Caguas

Population: 3,500,000

Area: 3,515 sq mi

Nicknames: Isle of Enchantment, Pearl of the Sea

Languages: Spanish, English

Mountain Ranges: Cordillera Central, Sierra Bermeja

Rivers: Rio de la Plata, Rio Grande de Loiza, Rio Portugues

Lakes and Reservoirs: Tortuguero Lagoon (Laguna Tortuguero), Dos Bocas Lake

Bays and Gulfs: San Juan Bay, Bioluminescent (Mosquito) Bay

Seas: Caribbean Sea

Straits: Mona Passage

Islands: Puerto Rico Island, Culebra Island, Vieques Island, Cayo Norte, Isla de Mona, Desecheo Island

Archipelagoes: Culebra Archipelago, Spanish Virgin Islands

Caves: Camuy River Cave System, Cueva Ventana

Other Bodies of Water: Ensenada Honda (Deep Cove), Condado Lagoon

Mountains: Cerro de Punta, El Yunque, Monte Jayuya, Cerro Rosa, Cerro Maravilla

National Forests: El Yunque National Forest

Other Sites: Castillo San Felipe del Morro, Porta Coeli (Gateway to Heaven), Arecibo Observatory, Castillo San Cristobal (Fort San Cristobal), San Juan National Historic Site, Flamenco Beach, Culebra National Wildlife Refuge, Old San Juan

Dams: Carraizo Dam, Portugues Dam

International/Major Airports: Luis Munoz Marin International Airport, Rafael Hernandez International Airport, Mercedita International Airport

U.S. VIRGIN ISLANDS

Capital: Charlotte Amalie

Major Cities: Charlotte Amalie, Christiansted

Population: 110,000

Area: 134 sq mi

Languages: English, English Creole, Spanish, French, French Creole

Rivers: Salt River

Bays and Gulfs: Magens Bay, Trunk Bay, Salt River Bay

Seas: Caribbean Sea

Straits: Leeward Passage

Islands: Saint Croix, Saint Thomas, Saint John, Water Island, Hassel Island, Hans Lollik Island, Flanagan Island, Buck Island, Capella Island

Archipelagoes: Virgin Islands

Mountains: Crown Mountain, Mount Eagle, Bordeaux Mountain

National Parks: Virgin Islands National Park

National Monuments: Virgin Islands Coral Reef National Monument, Buck Island Reef National Monument

Other Sites: Christiansted National Historic Site, Salt River Bay National Historical Park and Ecological Preserve

International/Major Airports: Cyril E. King Airport, Henry E. Rohlsen Airport

AMERICAN SAMOA

Capital: Pago Pago

Major Cities: Pago Pago, Tafuna, Nu'uuli

Population: 55,000

Area: 77 sq mi

Languages: Samoan, English, Tongan

Bays and Gulfs: Fagatele Bay, Faga'itua Bay

Peninsulas, Capes, and Points: Cape Taputapu

Islands: Tutuila, Ta'u, Ofu-Olosega Twin Islands, Aunu'u, Swains Island (Olosega Island)

Archipelagoes: Manu'a Islands

Other Landforms: Vailulu'u

Other Bodies of Water: Pago Pago Harbor

Mountains: Lata Mountain, Rainmaker Mountain (North Pioa Mountain), Matafao Peak

National Parks: National Park of American Samoa

National Monuments: Rose Atoll Marine National Monument

Other Sites: Fagatele Bay National Marine Sanctuary, Rose Island Concrete Monument

International/Major Airports: Pago Pago International Airport

GUAM

Capital: Hagatna

Major Cities: Dededo, Hagatna

Population: 165,000

Area: 210 sq mi

Languages: Chamorro, English

Rivers: Talofofo River, Pago River

Lakes and Reservoirs: Fena Lake

Bays and Gulfs: Agat Bay, Talofofo Bay

Seas: Philippine Sea

Peninsulas, Capes, and Points: Orote Peninsula, Point Udall, Ritidian Point, Pati Point, Aga Point, Facpi Point, Paseo de Susana Peninsula

Islands: Guam, Cocos Island, Cabras Island, Fofos Island, Agrigan Island

Archipelagoes: Mariana Islands

Other Landforms: Uruno Beach, Fouha Point

Other Bodies of Water: Cocos Lagoon, Apra Harbor, Talofofo Falls

Mountains: Mount Lamlam, Mount Jumullong Manglo, Mount Bolanos,

Other Sites: Naval Base Guam, War in the Pacific National Historical Park, Asan Invasion Beach, Fort Nuestra Senora de la Soledad, Dulce Nombre de Maria Cathedral Basilica, UnderWater World Guam

International/Major Airports: Antonio B. Won Pat International Airport

NORTHERN MARIANA ISLANDS

Capital: Saipan

Major Cities: Saipan

Population: 55,000

Area: 179 sq mi

Languages: Chamorro, Carolinian, English

Bays and Gulfs: Laolao Bay

Seas: Philippine Sea

Straits: Saipan Channel, Tinian Channel

Islands: Saipan Island, Rota Island, Tinian Island, Managaha Island, Anatahan Island, Pagan Island, Agrihan Island, Asuncion Island, Farallon de Parajos, Guguan Island, Alamagan Island, Sarigan Island, Farallon de Medinila, Aguigan Island

Archipelagoes: Northern Mariana Islands, Mariana Islands, Maug Islands

Other Landforms: Marpi Reef, Zealandia Bank

Other Bodies of Water: Tanapag Harbor

Mountains: Mount Agrihan, Mount Tapochau, Anatahan Volcano

National Monuments: Marianas Trench Marine National Monument

Other Sites: American Memorial Park, The Grotto

International/Major Airports: Saipan International Airport, Rota International Airport

WASHINGTON, D.C. (DISTRICT OF COLUMBIA)

Population: 695,000

Area: 68 sq mi

Rivers: Potomac River, Anacostia River, Rock Creek

Lakes and Reservoirs: Tidal Basin, Kingman Lake, McMillan Reservoir, Dalecarlia Reservoir, Georgetown Reservoir

Plains and Basins: Atlantic Coastal Plain

Peninsulas, Capes, and Points: Buzzard Point (Arsenal Point)

Islands: Theodore Roosevelt Island, Columbia Island, Three Sisters, Hains Point

Other Bodies of Water: Washington Channel, Chesapeake and Ohio Canal

Other Sites: United States Capitol, White House, Supreme Court Building, Washington Monument, Lincoln Memorial, Smithsonian Institution Building, Jefferson Memorial, Martin Luther King Jr. Memorial, Washington National Cathedral, Vietnam Veterans Memorial, Eisenhower Executive Office Building, Lincoln Memorial Reflecting Pool, National Mall, National Air and Space Museum, Jefferson Pier, Franklin Delano Roosevelt Memorial, George Mason Memorial, District of Columbia War Memorial, Library of Congress, National Museum of Natural History, National Museum of American History, National Museum of African Art, National Museum of the American Indian (Native Americans), Smithsonian American Art Museum, National Zoo, National Gallery of Art, National Geographic Society Museum, International Spy Museum, Hirshhorn Museum and Sculpture Garden, , National Portrait Gallery, Renwick Gallery, Union Station, Judiciary Square, National Building Museum, United States Holocaust Memorial Museum, John F. Kennedy Center for the Performing Arts, Lincoln Theater, Howard

Theater, Ford's Theater, William H.G. FitzGerald Tennis Center, Nationals Park, Georgetown University, Howard University, Lafayette Square, The Ellipse (President's Park South), Theodore Roosevelt Island National Memorial, Japanese American Memorial to Patriotism During World War II

International/Major Airports: Baltimore-Washington International Airport, Ronald Reagan Washington National Airport, Washington Dulles International Airport

Geographic Extremes of the United States

Northernmost Point: Point Barrow, Alaska

Northernmost Incorporated Place: Barrow, Alaska

Northernmost Point in the Contiguous United States: Northwest Angle, Lake of the Woods, Minnesota

Northernmost Incorporated Place in the Contiguous United States: Sumas, Washington

Southernmost Point: Ka Lae (South Point), Hawaii

Southernmost Incorporated Place: Honolulu County, Hawaii

Southernmost Point in the Contiguous United States: Cape Sable, Florida

Southernmost Incorporated Place in the Contiguous United States: Key West, Florida

Easternmost Point: Semisopochnoi Island, Aleutian Islands, Alaska

Easternmost Incorporated Point: Lubec, Maine

Easternmost Point in the Contiguous United States: West Quoddy Head, Maine

Easternmost Incorporated Place in the Contiguous United States: Lubec, Maine

Westernmost Point: Amatignak Island, Alaska

Westernmost Incorporated Place: Adak, Alaska

Westernmost Point in the Contiguous United States: Cape Alava, Washington

Westernmost Incorporated Place in the Contiguous United States: Port Orford, Oregon

Highest Point in the United States: Denali, Alaska

Highest Point in the Contiguous United States: Mount Whitney, California

Highest Point on the North American Continental Divide: Mount Elbert, Colorado

Highest Island Point in the United States: Mauna Kea, Hawaii

Highest City in the United States: Leadville, Colorado

Highest Town in the United States: Alma, Colorado

Highest County in the United States: San Juan County, Colorado

Highest U.S. State by Average Elevation: Colorado

Highest Point East of the Mississippi River: Mount Mitchell

Highest Airfield in the United States: Lake County Airport

Lowest Point in the United States: Badwater Basin, Death Valley, California

Lowest Lake in the United States: Salton Sea, California

Lowest High Point in the United States: Britton Hill, Florida

Lowest City in the United States: Calipatria, California

Lowest U.S. State by Average Elevation: Delaware

Lowest Airfield in the United States: Furnace Creek Airport, California

Largest Island in the United States: Hawaii

Longest Barrier Island in the United States: Padre Island

Largest Island in a Lake in the United States: Isle Royale

Largest Lake in the United States: Lake Superior

Longest River in the United States: Missouri River

Longest River System in the United States: Mississippi-Missouri River System

About the Author

Keshav Ramesh is a 14-year old author of twenty books, including *The Geography Bee Ultimate Preparation Guide*, *A Competitor's Compendium to the Geography Bee*, *The Quintessential Questionnaire to the Geography Bee*, and *The Geography Bee Comprehensive U.S. Reference Guide*.

These geography bee books have been designed to prepare students for all levels of the National Geographic Bee (NGB), United States Geography Olympiad (USGO), International Geography Bee (IGB), and North South Foundation (NSF) Junior/Senior Geography Bees.

Keshav competed in the CT Geographic Bee for three years and ranked 16th nationally in the 2016 NSF Senior GeoBee National Competition in Tampa, Florida. He has also taken the USGO/IGB NQEs several times. In addition, Keshav has competed in geology-related events at the CT Science Olympiad competition.

Keshav lives in Connecticut with his family.

Bibliography

"Geography." *ThoughtCo.* About, Inc., n.d. Web. <https://www.thoughtco.com/geography-4133035>.

National Geographic. *National Geographic Atlas of the World, Tenth Edition.* 10th ed. N.p.: National Geographic Society, 2014. Print.

National Geographic. *National Geographic Kids United States Atlas.* N.p.: National Geographic Society, 2012. Print.

National Geographic Kids. *National Geographic United States Encyclopedia: America's People, Places, and Events.* N.p.: National Geographic Society, 2015. Print.

Wikipedia. Wikimedia Foundation, n.d. Web. <https://www.wikipedia.org/>.

Wojtanik, Andrew. *National Geographic Bee Ultimate Fact Book: Countries A to Z.* N.p.: National Geographic Society, 2012. Print.

38387810R00100

Made in the USA
San Bernardino, CA
10 June 2019